THE MARINE
SEXTANT

THE MARINE SEXTANT

Selected from

AMERICAN PRACTICAL
NAVIGATOR

Nathaniel Bowditch

David McKay Company, Inc.

NEW YORK

THE MARINE SEXTANT

Selected from

AMERICAN PRACTICAL NAVIGATOR

Published by
David McKay Company, Inc.
750 Third Avenue
New York, New York 10017

No portion of this edition
may be mechanically reproduced by any means
without approval in writing from
the publisher.

1976

Library of Congress Cataloging in Publication Data

Bowditch, Nathaniel, 1773-1838.
    The marine sextant.

    1. Sextant.   2.   Nautical astronomy.   I.   American
practical navigator.   II.   Title.
VK583.B73   1976   527′.028   76-41052
ISBN 0-679-50653-5
ISBN 0-679-50668-3 pbk.

MANUFACTURED IN THE UNITED STATES OF AMERICA

The editors have included a comparison of commercial nautical sextants as an appendix to this volume. This material was not included in the original edition of AMERICAN PRACTICAL NAVIGATOR.

Contents

THE MARINE
SEXTANT

INSTRUMENTS FOR
CELESTIAL NAVIGATION

The Marine Sextant

The marine sextant is a hand-held instrument for measuring the angle between the lines of sight to two points by bringing into coincidence at the eye of the observer the direct ray from one point, and a double-reflected ray from the other, the measured angle being twice the angle between the reflecting surfaces.

The sextant's principal use is to measure the altitudes of celestial bodies above the visible sea horizon. Sometimes it is turned on its side and used for measuring the *difference* in bearing of two terrestrial objects. Because of its great value for determining position at sea, the sextant has been a symbol of navigation for more than 200 years. The quality of his instrument, the care he gives it, and the skill with which he makes observations are to the navigator matters of professional pride.

The name "sextant" is from the Latin *sextans,* "the sixth part." The arc of early marine sextants was approximately the sixth part of a circle, but because of the optical principle involved, the instrument measured angles of 120°. Most modern instruments measure something more than this.

Principle of operation—When a ray of light is reflected from a plane surface, the **angle of reflection** is equal to the **angle of incidence** (fig. 1a). When the reflecting surface is rotated toward or away from the incident ray, each angle is

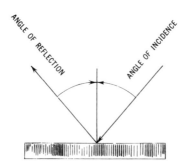

FIGURE 1a—*Angle of reflection equals angle of incidence.*

changed by the amount of rotation, so that the total angle between the incident and reflected rays is altered by twice the change in the reflecting surface. With the sextant, the ray of light is reflected by two mirrors; one movable and the other fixed. The angle between the first and last directions is twice the angle between the mirrors.

In figure 1b, *AB* is a ray of light from a celestial body. The index mirror of the sextant is at *B*, the horizon glass at *C*, and the eye of the observer at *D*. Construction lines *EF*

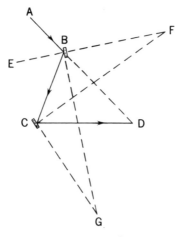

FIGURE 1b—*Optical principle of the marine sextant.*

and *CF* are perpendicular to the index mirror and horizon glass, respectively, and lines *BG* and *CG* are parallel to these mirrors. Therefore, angles *BFC* and *BGC* are equal because their sides are mutually perpendicular. Angle *BGC* is the inclination of the two reflecting surfaces. The ray of light *AB* is reflected at mirror *B*, proceeds to mirror *C*, where it is again reflected, and then continues on to the eye of the observer at *D*. Since the angle of reflection is equal to the angle of incidence,

$$ABE = EBC, \text{ and } ABC = 2EBC$$
$$BCF = FCD, \text{ and } BCD = 2BCF.$$

Since an exterior angle of a triangle equals the sum of the two nonadjacent interior angles

$$ABC = BDC + BCD, \text{ and } EBC = BFC + BCF.$$

Transposing,

$$BDC = ABC\text{-}BCD, \text{ and } BFC = EBC\text{-}BCF.$$

Substituting *2EBC* for *ABC*, and *2BCF* for *BCD* in the first of these equations,

$$BDC = 2EBC\text{-}2BCF, \text{ or } BDC = 2(EBC\text{-}BCF).$$

Since

$$BFC = EBC\text{-}BCF, \text{ and } BFC = BGC,$$

therefore

$$BDC = 2BFC = 2BGC.$$

That is, *BDC*, the angle between the first and last directions of the ray of light, is equal to *2BGC*, twice the angle of inclination of the reflecting surfaces. Angle *BDC* is the altitude of the celestial body.

Micrometer drum sextant—A modern marine sextant,

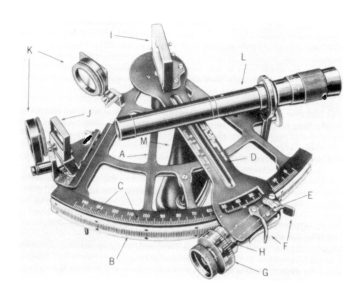

FIGURE 2—*U. S. Navy standard micrometer drum sextant.*

called a **micrometer drum sextant,** is shown in figure 2. In most marine sextants, the **frame,** *A,* is made of brass or aluminum. There are several variations of the design of the frame, nearly all conforming generally to that shown. The **limb,** *B,* is cut on its outer edge with teeth, each representing one degree of altitude. The altitude graduations, *C,* along the limb, are called the **arc.** Some sextants have an arc marked in a strip of brass, silver, or platinum inlaid in the limb.

The **index arm,** *D,* is a movable bar of the same material as the frame. It is pivoted about the center of curvature of the limb. The **tangent screw,** *E,* is mounted perpendicularly on the end of the index arm, where it engages the teeth of the limb. Because the index arm can be moved through the length of the arc by rotating the tangent screw, this is sometimes called an "endless tangent screw," in contrast with the limited-range device on older instruments. The **release,** *F,* is a spring-actuated clamp which keeps the tangent screw engaged with the teeth of the limb. By applying pressure on the legs of the release, one can disengage the tangent screw. The index arm can then be moved rapidly along the limb. Mounted on the end of the tangent screw is a **micrometer drum,** *G,* which is graduated in minutes of altitude. One complete turn of the drum moves the index arm one degree of altitude along the arc.

Adjacent to the micrometer drum and fixed on the index arm is a **vernier,** *H,* which aids in reading fractions of a minute. The vernier shown is graduated into ten parts, permitting readings to six seconds. Other sextants (generally of European manufacture) have verniers graduated into only six parts, permitting readings to ten seconds. The most expensive sextant in common use has no vernier, and readings more precise than one minute can only be estimated.

The **index mirror,** *I,* is a piece of silvered plate glass mounted on the index arm, perpendicular to the plane of the instrument, with the center of the reflecting surface directly over the pivot of the index arm. The **horizon glass,** *J,* is a piece of plate glass silvered on its half nearer the frame. It is mounted on the frame, perpendicular to the plane of the sextant. The index mirror and horizon glass are mounted so that their surfaces are parallel when the micrometer drum is set at 0°, if the instrument is in perfect adjustment.

Shade glasses, *K,* of varying or variable darkness, are mounted on the frame of the sextant in front of the index mirror and horizon glass. They can be moved into the line of sight at will, to reduce the intensity of light reaching the eye of the observer. Older sextants have two sets of shade glasses, as shown in figure 4. Many modern sextants are fitted with a single Polaroid **filter** of variable darkness in place of each set of shade glasses, as shown in figure 2.

The **telescope,** *L,* screws into an adjustable collar in line with the horizon glass, and should then be parallel to the plane of the instrument. Most modern sextants are provided with only one telescope, but some are equipped with two or more. When only one telescope is provided, it is of the "erect image type," either such as shown or one with a wider "object glass" (far end of telescope), which generally is shorter in length and gives a greater field of view. The second telescope, if provided, is of the "inverting type." The inverting telescope, having one lens less than the erect type, absorbs less light, but at the expense of producing an inverted image. A small colored glass cap is usually provided, to be placed over the "eyepiece" (near end of telescope) to reduce the glare. With this in place, shade glasses are generally not needed. A "peep sight" may be provided. It is a clear tube which serves to direct the line of sight of the observer when no telescope is used.

The telescope shown in figure 2 is fitted with a "spiral focusing mechanism." Other sextants substitute a "draw" for this mechanism. The draw is fitted inside the telescope tube without threads and is slid in or out as necessary to focus the instrument. The spiral focusing mechanism is easily adjusted each time the sextant is used, but on the draw type, the navigator should mark the draw to indicate the correct extension for his eyes.

The **handle,** *M,* of most sextants is made of wood or plastic. Sextants are designed to be held in the right hand. Some are equipped with a small light on the index arm to assist in reading altitudes. The batteries for this light are fitted inside a recess in the sextant handle.

Figure 3 shows a sextant with a silver arc inserted in a limb, a micrometer drum graduated oppositely to the one in figure 2, a vernier graduated into six parts, a shorter telescope with a wider object glass than that in figure 2, a telescope draw substituted for a spiral focusing mechanism, and a light fitted on the index arm.

Vernier sextant—Nearly all marine sextants of recent manufacture are of the type described earlier. At least two older-type sextants are still in use. These differ from the micrometer drum sextant principally in the manner in which the final reading is made. They are called **vernier sextants.**

Figure 3—*A micrometer drum sextant used in the merchant marine.*

The **clamp screw vernier sextant** is the older of the two. In place of the modern "release," a **clamp screw** is fitted on the underside of the index arm. To move the index arm, one loosens the clamp screw, releasing the arm. When the arm is placed at the approximate altitude of the body being observed, the clamp screw is tightened. Fixed to the clamp screw and engaged with the index arm is a long tangent screw. When this screw is turned, the index arm moves slowly, permitting accurate setting. Movement of the index arm (by the tangent screw) is limited to the length of the screw (several degrees of arc). Before an altitude is measured, this screw should be set to the

FIGURE 4—*A clamp screw vernier sextant.*

approximate mid-point of its range. The final reading is made on a vernier set in the index arm below the arc. A small microscope or magnifying glass fitted to the index arm is used in making the final reading. Figure 4 shows a clamp screw vernier sextant.

The **endless tangent screw vernier sextant** is identical with the micrometer drum sextant, except that it has no drum, and the fine reading is made by a vernier along the arc, as with the clamp screw vernier sextant. The release is the same as on the micrometer drum sextant and teeth are cut into the underside of the limb which engage with the endless tangent screw. The vernier itself is explained later in this section.

Use of the Sextant

Use of the sextant—When the *sun* is observed, the sextant is held vertically in the right hand, and the line of sight is directed at the point on the horizon directly below the body. Suitable shade glasses are moved into the line of sight, and the index arm is moved outward from near the 0° point until the reflected image of the sun appears in the horizon glass, near the direct view of the horizon. The sextant is then tilted slightly to the right and left to check its perpendicularity. As the sextant is tilted, the image of the sun appears to move in an arc, and the observer may have to change slightly the direction in which he is facing, to prevent the image from moving out of the horizon glass. When the sun appears at the *bottom* of its apparent arc resulting from this **swinging the arc,** or **rocking the sextant,** the sextant is vertical, and in the correct position for making the observation. If the sextant is tilted, too *great* an angle will be measured. When the sextant is vertical, and the observer is facing directly toward the sun, its reflected image appears at the center of the horizon glass, half on the silvered part, and half on the clear part. The index arm is then moved slowly until the sun appears to be resting exactly on the horizon, which is tangent to the **lower limb.**

Occasionally, the sun image is brought *below* the horizon, and the **upper limb** observed. It is good practice to make several observations, moving the limb away from the horizon, alternately above and below it, between readings. Practice is needed to determine the appearance at tangency, which occurs at only one point, to avoid the common error of beginners of bringing the image down too far (too little for an upper-limb observation).

Some navigators get more accurate observations by letting the body contact the horizon by its own apparent motion, bringing it slightly below the horizon if rising, and

FIGURE 5—TOP LEFT, *view through telescope with index arm set near zero.* TOP RIGHT, *"swinging the arc" after the sun has been brought close to the horizon.* BOTTOM, *sun at the instant of tangency.*

above if setting. At the instant the horizon is tangent to the disk, the time is noted.

The **sextant altitude** is the uncorrected reading of the sextant. Figure 5 illustrates the major steps in making an observation of the sun. At the left, the index arm has been moved a short distance from 0°. In the center, it has been clamped with the sun in the approximate position for a reading, and the sextant is being rocked. At the right, the sun is in the correct position for a reading.

When the *moon* is observed, the procedure is the same as for the sun, except that shade glasses are usually not required. The upper limb of the moon is observed more often than that of the sun, because of the phases of the moon. When the terminator—the line separating the light and dark portions—is nearly vertical, care should be exercised in selecting the limb that is illuminated, if an inaccurate reading is to be avoided. Sights of the moon are best made during daylight hours, or during that part of twilight in which the moon is least luminous. During the night, false horizons nearly always appear below the moon, due to illumination of the water by moonlight.

When a *star* or *planet* is observed, three methods of making the initial approximation of the altitude are in common use. In the most common, the index arm and micrometer drum are set on zero and the line of sight is directed at the body to be observed. Then, while keeping the reflected image of the body in the mirrored half of the horizon glass, the index arm is slowly swung *out* and the frame of the sextant is rotated *down*. The reflected image of the body is kept in the mirror until the horizon appears in the clear part of the horizon glass.

When there is little contrast between brightness of the sky and the body, this procedure is difficult, for if the body is "lost" while it is being brought down, it may not be recovered without starting again at the beginning of the procedure.

An alternative method sometimes used consists of holding the sextant upside down in the left hand, directing the line of sight at the body, and slowly moving the index arm out until the horizon appears in the horizon glass. This is illustrated in figure 6. After contact is made, the sextant is inverted and the sight taken in the usual manner.

A third method consists of determining in advance the approximate altitude and azimuth of the body by a **star finder** such as H.O. 2102-D, below. The sextant is set at the indicated altitude, and the observer faces in the direction indicated by the azimuth. After a short search, during which the index arm is moved backward and forward a few degrees, and the azimuth in which the observer faces is changed a little to each side, the image of the body should appear in the horizon glass. The best method to use for any observation is that which produces the desired result

11

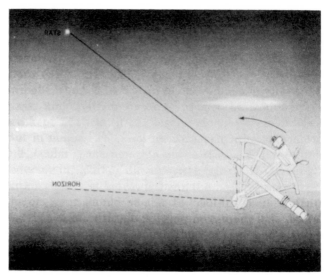

FIGURE 6—*Method of bringing horizon "up" to body.*

with the least effort. It is largely a matter of personal preference.

Measurement of the altitude of a star or planet differs from that of the sun or moon in that the *center* of a star or planet, rather than a limb, is brought into coincidence with the horizon. Figure 7 shows the reflected image of a star as it should appear at the time of observation. Because of this difference, and the limited time usually available for observation during twilight, the method of letting a star or planet intersect the horizon by its own motion is little used. As with the sun and moon, however, the navigator should not forget to swing the arc to establish perpendicularity of the sextant.

Occasionally, fog, haze, or other ships may obscure the horizon directly below a body which the navigator wishes to observe. If the arc of the sextant is sufficiently long, a **back sight** might be obtained, using the opposite point of the horizon as the reference. The observer faces *away* from

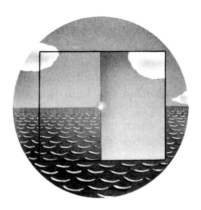

FIGURE 7—*Correct position of planet or star at moment of observation.*

the body and observes the *supplement* of the altitude. If the sun or moon is observed in this manner, what appears in the horizon glass to be the lower limb is in fact the upper limb. In the case of the sun, it is usually preferable to observe what appears to be the upper limb. The arc that appears when rocking the sextant for a back sight is inverted; that is, the *highest* point indicates the position of perpendicularity.

If more than one telescope is furnished with the sextant, the erecting telescope is used to observe the sun. Generally, the inverting telescope will produce the best results when observing the stars, although some navigators prefer not to use any telescope, thus obtaining a wider field of view. The collar into which the sextant telescope fits may be adjusted in or out in relation to the frame. When moved in, more of the mirrored half of the horizon glass is visible to the navigator, and a star or planet is more easily observed when the sky is relatively bright. Near the darker limit of twilight, the telescope can be moved out, giving a broader view of the clear half of the glass, and making the less distinct horizon more easily discernible. If both eyes are kept open until the last moments of an observation, eye strain will be lessened. But in making the final measurement, the nonsighting eye should be closed to permit full ocular concentration.

Practice will permit observations to be made quickly, reducing inaccuracy due to eye fatigue. If several observations are made in succession, with a short rest between them, the best results should be obtained. With experience, the observer should be able to "call his shots," identifying the better ones.

When an altitude is being measured, it is desirable to have an assistant note the time, so that simultaneous values of time and altitude will be available. He should be given a warning "stand-by" when the measurement is nearly completed, and a "mark" at the moment a reading is made. He should be instructed to read the three hands in order of their rapidity of motion; the second hand first, then the minute hand, and finally the hour hand. If it is sufficiently dark that a light is needed to make the reading, the assistant should read both the time, and then the altitude, *behind* the observer and facing away from him, to avoid impairment of the observer's eye adaption to sky and horizon lighting conditions.

If an assistant is not available to time the observations, the observer holds the watch in the palm of his left hand, leaving his fingers free to manipulate the tangent screw of the sextant. After making the observation, he quickly shifts his view to the watch, and notes the positions of the second, minute, and hour hands, respectively. The delay between completing the altitude observation and noting the time should not be more than one or two seconds. The average time should be determined by having someone measure it for several observations, or by counting the half seconds (learning to count with the half-second beats of a chronometer). This interval can then be subtracted from the observed time of each sight.

Reading the Sextant

Reading the sextant—The reading of a micrometer drum sextant is made in three steps. The degrees are read by noting the position of the arrow on the index arm in relation to the arc. The minutes are read by noting the position of the zero on the vernier with relation to the graduations on the micrometer drum. The fraction of a minute is read by noting which mark on the vernier most nearly coincides with one of the graduations on the micrometer drum. This is similar to reading the time by means of the hour, minute, and second hands of a watch. In both, the relationship of one part of the reading to the others should be kept in mind. Thus, if the hour hand of a watch were *about* on "4," one would know that the time was about four o'clock. But if the minute hand were on "58," one would know that the time was 0358 (or 1558), not 0458 (or 1658). Similarly, if the arc indicated a reading of about 40°, and 58′ on the micrometer drum were opposite zero on the vernier, one would know that the reading was 39°58′, not 40°58′. Similarly, any doubt as to the correct minute can be removed by noting the fraction of a minute from the position of the vernier. In figure 8 the reading is 29°42′.5. The arrow on the index mark is between 29° and 30°, the zero on the vernier is between 42′ and 43′, and the "0′.5" graduation on the vernier coincides with one of the graduations on the micrometer drum.

The principle of reading a vernier type sextant is the same, but the reading is made in two steps. Figure 9 shows a typical altitude setting on this type sextant. Each degree on the arc of this sextant is graduated into three parts, permitting an initial reading by the reference mark on the

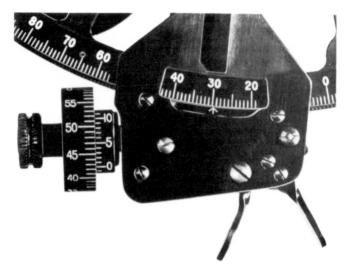

FIGURE 8—*Micrometer drum sextant set at 29°42'.5.*

index arm to the nearest full 20 minutes of arc. In this illustration the reference mark lies between 29°40' and 30°00', indicating a reading between these values. The reading for the fraction of 20' is made by means of the vernier, which is engraved on the index arm and has the small reference mark as its zero graduation. On this vernier, 40 graduations coincide with 39 graduations on the arc. Each graduation on the vernier is equivalent to 1/40 of one graduation (20') on the arc, or 0.'5 (30"). In the illustration, the vernier graduation representing 2½ minutes (2'30") most nearly coincides with one of the graduations on the arc. Therefore, the reading is 29°42'30", or 29°42.'5, as before. When a vernier of this type is used, any doubt as to which mark on the vernier coincides with a graduation on the arc can usually be resolved by noting the position of the vernier mark on each side of the one that seems to be in coincidence.

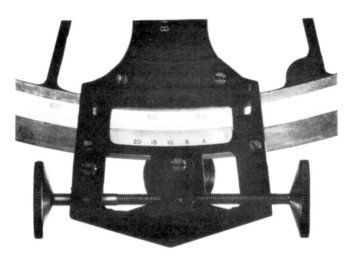

FIGURE 9—*Vernier sextant set at 29°42′30″.*

Developing Observational Skill

Developing observational skill—A well-constructed marine sextant is capable of measuring angles with an instrument error not exceeding 0′.1. Lines of position from altitudes of this accuracy would not be in error by more than about 200 yards. However, there are various sources of error, other than instrumental, in altitudes measured by sextant. One of the principal sources is the observer himself. There is probably no single part of his work that the navigator regards with the same degree of professional pride as his ability to make good celestial observations. Probably none of his other tasks requires the same degree of skill.

The first fix a student navigator obtains by his observation of celestial bodies is likely to be disappointing. Most navigators require a great amount of practice to develop the skill needed to make good observations. But practice alone is not sufficient, for if a mistake is repeated many times, it will be difficult to eradicate. Early in his career a navigator would do well to establish good observational technique—and continue to develop it during the remainder of his days as navigator. Many good pointers can be obtained from experienced navigators, but it should be remembered that each develops his own technique, and a practice that proves highly successful for one observer may not help another. Also, an experienced navigator is not necessarily a good observer, although he may consider himself such. Navigators have a natural tendency to judge the accuracy of their observations by the size of the figure formed when the lines of position are plotted. Although this is some indication, it is an imperfect one, because it does not indicate the errors of individual observations, and may not reflect constant errors. Also, it is a compound of a number of errors, some of which are not subject to control by the navigator.

When a beginner first begins to use the sextant, he can eliminate gross errors of principle in its use, and gain some ability in making observations, by accepting the coaching of an experienced navigator. By watching the novice make observations, the experienced navigator can observe a tendency to hold the instrument incorrectly, swing the arc improperly, or make other mistakes. When a celestial body is near the celestial meridian, the experienced navigator might make an observation and quickly transfer the sextant to the inexperienced one, who can see how the sight should appear. The two might make simultaneous observations and compare results. At first it is well to select bodies of low altitude, if they are available.

This procedure is helpful in detecting gross mistakes, but since the observations of the experienced navigator are not without error, this method is not suitable for final polishing of technique. For this purpose, observations should be compared with a more exact standard. Lines of position from celestial observations can be compared with good positions obtained by electronics or by piloting, if near a shore. Although this is good practice and provides a means of checking one's skill from time to time, it does not provide the large number of comparisons in a short time needed if technique is to be perfected.

This can sometimes be accomplished when a vessel is at anchor, or at a pier, if a stretch of open horizon is available. In advance, the altitude of a celestial body which will be over the open horizon at a time favorable for observation is computed at intervals of perhaps eight minutes (change in hour angle of 2°). If the body will be near the meridian, a smaller interval should be used. The altitude is determined for the position of the vessel, and all sextant altitude corrections (see below) are applied with reversed sign. These altitudes are then plotted versus time on cross-section paper, to a large scale, and a curve drawn through the points. At the selected time, a large number of observations are made at short intervals, allowing only enough time between observations for resting the eyes and arms. These observations are then plotted on the cross-section paper and compared with the curve.

An analysis of the results should be instructive. Erratic results indicate poor observational conditions or the need for practice and more care in making observations.

If the measured altitudes are consistently too great, the sextant may not be rocked properly, the condition of tangency of the lower limb of the sun or moon may not be judged accurately, a false horizon in the water may have been used, subnormal refraction (dip) might be present, the eye might be higher above water than estimated, time might be in error, the index correction may have been determined incorrectly, the sextant might be out of adjustment, an error may have been made in the computation, the horizontal (vertical) may be tilted slightly by nearby mountains, etc.

If the measured altitudes are consistently too low, the condition of tangency of the upper limb of the sun or moon may not be judged accurately, a low cloud may have been used as the horizon, abnormal refraction (dip) might be present, height of eye might be lower than estimated, time might be in error, the index error may have been determined incorrectly, the sextant might be out of adjustment, an error may have been made in the computation, the waves or swell at the horizon might be higher than at the ship, the horizontal (vertical) may be tilted slightly, a planet or bright star may have been placed "tangent" to the horizon rather than centered on the horizon, etc.

A single test of this type, while instructive, may not be conclusive. Several tests should be made with different celestial bodies, at various altitudes, under various conditions of weather and sea, and at different places. Generally, it is possible and desirable to correct any errors being made in the technique of observation, but occasionally a **personal error** (sometimes called **personal equa-**

tion) will persist. This might be different for the sun and moon than for planets and stars, and might vary with degree of fatigue of the observer, and other factors. For this reason, a personal error should be applied with caution. However, if a relatively constant personal error persists, and experience indicates that observations are improved by applying a correction to remove its effect, better results might be obtained by this procedure than by attempting to eliminate it from one's observations.

When lines of position of great reliability are desired, even an experienced navigator can usually improve his results by averaging to reduce random error. A number of observations, preferably not less than ten, are made in quick succession. These can then be plotted versus time, on cross-section paper, and a curve faired through the points. Unless the body is near the celestial meridian, this curve should be very nearly a straight line. *Any* point on the curve can be used as the observation, using the time and altitude indicated by the point. It is best to use a point near the middle of the line.

A somewhat simpler variation is generally available if observations are made at equal intervals, unless the body is near the meridian. It is based upon the assumption that the change in altitude should be equal for equal intervals of time. A number of observations might be made by having an assistant give a warning "stand-by" and then a "mark" at equal intervals of time, at every ten or 20 seconds. Perhaps a better procedure is to make the observations at equal altitude increments. After the first observation, the altitude is changed by a set amount according to its rate of change, as 5'. The setting is *increased* if the body is rising, and *decreased* if it is setting. The body is then permitted to cross the horizon by its own motion, and at the instant of doing so, the time is noted. If time intervals are constant, the *mid time* and the *average altitude* are used as the observation. If altitude increments are constant, the *average time* and *mid altitude* are used. An uneven number of observations simplifies the finding

21

of the mid value, but with ten observations the finding of the average value is easier.

If only a small number of observations is available, as three, it is usually preferable to solve all observations and plot the resulting lines of position, adjusting them to a common time. The *average* position of the line might be used, but it is generally better practice to use the middle line (or a line midway between the two middle ones if there are an even number).

In this discussion of averaging, it has been assumed that all observations are considered of nearly equal value. Any observation considered unreliable, either in the judgment of the observer or as a result of a plot, should be rejected in finding an average.

Need for sextant altitude corrections—Altitudes of celestial bodies, obtained aboard ship for the purpose of establishing lines of position, are normally measured by a hand-held **sextant.** The uncorrected reading of a sextant after such an operation is called **sextant altitude (hs)**. If the sextant is in proper adjustment, certain sources of error are eliminated. There remains, however, a number of sources of error over which the observer has little or no control. For each of these he applies a correction. When all of these **sextant altitude corrections** have been applied, the value obtained is the altitude of the center of the celestial body above the celestial horizon, for an observer at the center of the earth. This value, called **observed altitude (Ho)**, is compared with the **computed altitude (Hc)** to find the **altitude difference** (*a*) used in establishing a line of position.

Care of the Sextant

Care of the sextant—The modern marine sextant is a well-built precision instrument capable of rendering many years of reliable service, with minimum attention. However, its usefulness can easily be impaired by careless handling or neglect. If it is ever dropped, it may never again provide reliable information. If this occurs, the instrument should be taken to an expert for careful testing and inspection.

When not in use, a sextant should invariably be kept in its case and properly stowed. The sextant case should be a well-constructed hardwood box fitted on its exterior with a lock, a handle, and two hooks, preferably the type having safety catches. The interior of the case should be fitted with blocks in which the handle or legs, or both, are placed when the sextant is stowed. Some sextant cases are fitted with catches which clamp over the handle when the sextant is stowed, and some are fitted with felt-lined blocks on the inside of the cover, to clamp down on the extreme ends of the arc when the case is closed. The case should be so constructed that it can be closed with the shade glasses and index arm in nearly any normal position, and preferably with the telescope in place. The last is particularly valuable to the navigator on an overcast day when only one opportunity to observe the sun may present itself, and the sight may have to be taken quickly.

The case itself should be securely stowed in a convenient place away from excessive heat, dampness, and vibration. A shelf with built-up sides into which the case fits snugly is a good stowage place. The practice of leaving the sextant in its case on a chart room settee is a bad one, and the instrument should *never* be left unattended on the chart table.

To remove the sextant from its case, grasp the frame firmly with the left hand, making sure that no pressure is applied to the index arm, and lift the instrument from the box. Then take the sextant in the right hand, by its handle, leaving the left hand free to make any adjustments necessary before taking a sight. The instrument should never be held by its limb, index arm, or telescope.

Next to careless handling, the greatest enemy of the sextant is moisture. The mirrors, especially, and the arc should be wiped dry after each use. A new sheet of plain lens paper is best to use for this purpose, and linen second best. Over a period of time, however, linen collects dust, which may contain abrasives that will scratch the surface of the mirrors. For this reason, linen, if it is used, should be kept in a small bag to protect it from dust in the air. Chamois leather and silk are particularly likely to collect abrasive dusts from the air and they should not be used to clean the mirrors or telescope lenses. Should the mirrors become particularly dirty, they can be cleaned with a small amount of alcohol, applied with a clean piece of lens paper. The arc can be cleaned, when necessary, with ammonia, but never with a polishing compound. In cleaning or drying the mirrors and arc, care should be taken that excessive pressure is not applied to any part of the instrument.

A small bag of silica gel kept in the sextant case will help in keeping the air in the case free from moisture, and will help to preserve the mirrors. Occasionally, the silica gel should be heated in an oven to remove the absorbed moisture.

The tangent screw and the teeth on the side of the limb should be kept clean and lightly oiled, using the oil provided with the sextant. It is good practice to set occasionally the index arm of an endless tangent screw at one extremity of the limb and then to rotate the tangent screw over the length of the arc. This will clean the teeth and spread the oil through them. At any time that the sextant is to be stowed for a long period, the arc should be protected with a thin coat of petroleum jelly.

If the mirrors need resilvering, they are best taken to an instrument shop where a professional job can be done. However, on rare occasions it may be necessary to resilver the mirrors of a sextant at sea. In anticipation of this possibility, the navigator should obtain the necessary materials in advance, as makeshift substitutes cannot be relied upon to do the job adequately. The required materials are xylene (available in most pharmacies), dilute nitric acid (optional), alcohol, cotton, tin foil about 0.005 inch thick, a small amount of mercury, a clean blotter, and some tissue paper. Do *not* substitute aluminum foil commonly used in packaging candy and cigarettes!

First, remove the protective coating with alcohol (or better, acetone) from the back of the mirror to be resilvered, and clean the glass with xylene or acid. If the old silvering is difficult to remove, soak it in water. Place the blotter on a flat surface and turn up and seal the edges to form a tray. This will serve to contain the mercury if the vessel should roll during the operation. Using cotton, clean and smooth out both sides of a piece of tin foil slightly larger than the glass to be silvered, first with alcohol and then with xylene (do not use acid). Make certain that no lint adheres to the foil, and place it on the blotter. Clean the mercury by squeezing it through cheese cloth, and apply a drop to the foil. Carefully spread it over the surface with a finger, making sure that none of the mercury gets under the foil. Add a few more drops of mercury until the entire surface of the foil is covered and tacky. The mercury combines with some of the tin to form an amalgam. Place the chemically cleaned glass on a piece of clean tissue paper with the side to be silvered face down. Then place the glass and the paper on the amalgam. Apply slight pressure to the glass and withdraw the tissue paper. Following this, grasp the edge of the tin foil and lift it and the mirror from the blotter. Invert the glass and the tin foil and place in an inclined position, silvered side up. Five or six hours later any loose foil may be scraped from the sides of the mirror, and the following day a coat of commercial varnish or lacquer should be applied to the silvered surface. Should the mirrored half of the horizon glass require silvering, the clear half may be protected by a strip of cellulose or adhesive tape.

Sextant Errors

Sextant adjustments—There are at least seven sources of error in the marine sextant, three nonadjustable by the navigator, and four adjustable.

The **nonadjustable errors** are: "prismatic error," "graduation error," and "centering error."

The **prismatic error** is present if the two faces of the shade glasses and mirrors are not parallel. Error due to lack of parallelism in the shade glasses may be called **shade error.** Shade error in the shade glasses near the index mirror can be determined by comparison of an angle measured when a shade glass is in the line of sight with the same angle measured when the glass is not in the line of sight. In this manner, the error for each shade glass can be determined and recorded. If shade glasses are used in combination, their combined error should be determined separately. If additional shading is needed for the observations, use the colored telescope eyepiece cover. This does not introduce an error because direct and reflected rays are traveling together when they reach it, and are therefore affected equally by any lack of parallelism of its two sides.

Lack of parallelism of the two faces of the index mirror can be detected by carefully measuring a series of angles; then removing the index mirror, inverting it, and replacing it; and then measuring the same angles again. Half the difference is the prismatic error. After the index mirror has been inverted, it should be checked carefully for perpendicularity to the frame of the sextant, as explained below.

Lack of parallelism of the two faces of the horizon glass will appear as part of the index error, and so need not have separate attention. The same is true of prismatic error in the shade glasses located near the horizon glass, but unless index error is determined with the shade glasses in place, the measured index error will not be the correct value for the combined error.

Graduation errors occur in the arc, micrometer drum, and vernier of a sextant which is improperly cut or incorrectly calibrated. Normally, the navigator cannot determine whether the arc of a sextant is improperly cut, but the principle of the vernier makes it possible to determine the existence of graduation errors in the micrometer drum or vernier and is a useful guide in detecting a poorly made instrument. The first and last markings on any vernier should align perfectly with one less graduation on the adjacent micrometer drum. In figure 2, the vernier is graduated in ten units. When the zero point is aligned with any graduation on the micrometer drum, the "ten" graduation should be in perfect alignment with a micrometer graduation nine units greater than the one in line with zero on the vernier. In figure 3, the vernier is graduated in six units and should align perfectly with any two graduations five units apart on the micrometer.

Centering error results if the index arm is not pivoted at the exact center of curvature of the arc. It can be determined by measuring known angles, after the adjustable errors have been removed. Horizontal angles can be used by determining the accurate value by careful measurement with a theodolite, an instrument designed to measure precise horizontal and verticle angles. Obviously, this must be done by an expert. Several readings by both theodolite and sextant should minimize errors. An alternative method is to measure angles between the lines of sight to stars, comparing the measured angles with computed values. To minimize refraction errors, one should select stars at about the same altitude, and avoid stars near the horizon.

The same shade glasses, if any, used for determining or eliminating index error should be used for measuring centering error. The errors determined in this manner include any error due to faulty graduation, and prismatic error of the index mirror, unless corrections are applied for these errors. However, since all vary with the angle measured, they need not be separated. Usually, it is preferable to make a single correction table for all three errors, called **instrument error.** Customarily, such a table is determined by the manufacturer and attached to the inside cover of the sextant case. The sign of the error is reversed, so that the values given are for **instrument correction (I).**

The **adjustable errors** in the sextant are those related to *perpendicularity* of (1) the frame and the index mirror, and (2) the frame and the horizon glass, and *parallelism* of (3) the index mirror and horizon glass to each other at zero setting, and of (4) the telescope to the frame. Each of these errors, if it exists, can be removed from the sextant by careful adjustment. In making these adjustments, *never tighten one adjusting screw without first loosening the other screw which bears on the same surface.* The adjustments should be made in the order indicated.

The first adjustment is for **perpendicularity of the index mirror** to the frame of the sextant. To test for perpendicularity, place the index arm at about 35° on the arc, and hold the sextant on its side, with the index mirror "up" and toward the eye. Observe the direct and reflected views of the sextant arc, as illustrated in figure 10. If the two views do not appear to be joined in a straight line, the index

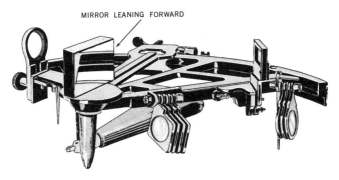

FIGURE 10—*Testing the perpendicularity of the index mirror. Here the mirror is not perpendicular.*

mirror is not perpendicular. If the reflected image is above the direct view, the mirror is inclined forward. If the reflected image is below the direct view, the mirror is inclined backward. An alternative and sometimes more satisfactory method of determining perpendicularity involves the use of two small vanes, or similar objects, of exactly the same height. Figure 11 illustrates this method.

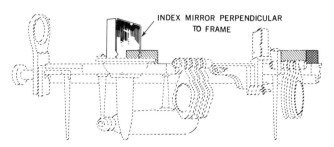

FIGURE 11—*Alternative method of testing the perpendicularity of the index mirror. Here the mirror is perpendicular.*

Again the index arm is set at about 35°. The vanes are placed upright on the extremities of the limb, in such a way that the observer can, by placing his eye near the index mirror, see the direct view of one vane and the reflected image of the other. The tops of the objects are then observed for alignment. The use of vanes permits observation in the plane of adjustment, rather than at an angle. Adjustment is made by means of two screws at the back of the index mirror.

The second adjustment is for **perpendicularity of the horizon glass** to the frame of the sextant. An error resulting from the horizon glass not being perpendicular is called **side error.** To test for perpendicularity, set the index arm at zero and direct the line of sight at a star. Then rotate the tangent screw back and forth so that the reflected image passes alternately above and below the direct view. If, in

30

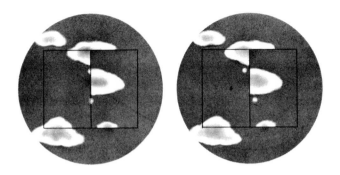

FIGURE 12—*Testing the perpendicularity of the horizon glass.* LEFT, *side error does not exist.* RIGHT, *side error does exist.*

changing from one position to the other, the reflected image passes directly over the star as seen without reflection, no side error exists, but if it passes to one side, the horizon glass is not perpendicular to the frame of the sextant. Figure 12 illustrates observations without side error (left) and with side error (right). Whether the sextant reads zero when the true and reflected images are in coincidence is immaterial in this test. An alternative method is to observe a vertical line, such as one edge of the mast of another vessel (or the sextant can be held on its side and the horizon used). If the direct and reflected portions do not form a continuous line, the horizon glass is not perpendicular to the frame of the sextant. A third method is to hold the sextant vertical, as in observing the altitude of a celestial body, and bring the reflected image of the horizon into coincidence with the direct view, so that it appears as a continuous line across the horizon glass. Then tilt the sextant right or left. If the horizon still appears continuous, the horizon glass is perpendicular to the frame, but if the reflected portion appears above or below that part seen direct, the glass is not perpendicular. Adjustment is made by means of two screws near the base of the horizon glass.

The third adjustment is to make the **index mirror and horizon glass parallel** when the index arm is set exactly at zero. The error which results when the two are not parallel is the principal cause of **index error,** the total error remaining after the four adjustments have been made. Index error should be determined each time the sextant is used and need not be removed if its value is known accurately. To make the test for parallelism of the mirrors, set the instrument at zero, and direct the line of sight at the horizon or a star. Side error having been eliminated, the direct view and reflected image of the horizon appear as a continuous line, or the star as a single point, if the two mirrors are parallel. If the mirrors are not parallel, the horizon appears broken at the edge of the mirrored part of the horizon glass, one part being higher than the other. The reflected image of a star appears above or below the star seen without reflection. If the star appears as a single point, move the tangent screw a small amount to be sure both direct view and reflected image are in the range of vision. The sun can be used by noting the reading when the reflected image is tangent to the sun as seen direct, first above it and then below. These should be numerically equal but of opposite sign (one positive and the other negative). To avoid variations in refraction, do not use low altitudes; or turn the sextant on its side and use the two sides of the sun. Adjustment is made by two screws near the base of the horizon glass. If the error is not to be removed, turn the tangent screw until direct view and reflected image of the horizon or a star are in coincidence. The reading of the sextant is the index error. It is positive if the reading is "on the arc" (positive angle), and negative if "off the arc" (negative angle). In the case of the sun it is *half* the numerical difference (algebraic sum) of the readings, positive or negative to agree with the larger reading.

Index correction (IC) is numerically the same as index error, but of opposite sign. Since both the second and third adjustments involve the position of the horizon glass, it is good practice to recheck for side error after index error has been eliminated. Index error should always be checked after adjustment for side error.

The fourth adjustment is to make the **telescope parallel** to the frame of the sextant. If the line of sight through the telescope is not parallel to the plane of the instrument, an **error of collimation** will result, and altitudes will be measured as greater than their actual values. To check for parallelism of the telescope, insert it in its collar, and observe two stars 90° or more apart, bringing the reflected image of one into coincidence with the direct view of the other, near either the right or left edge of the field of view (the upper or lower edge if the sextant is horizontal). Then tilt the sextant so that the stars appear near the opposite edge. If they remain in coincidence, the telescope is parallel to the frame, but if they separate, it is not. An alternative method is to place the telescope in its collar and then lay the sextant on a flat table. Sight along the frame of the sextant and have an assistant place a mark on the opposite bulkhead, in line with the frame. Place another mark above the first at a distance equal to the distance from the center of the telescope to the frame. This second line should be in the center of the field of view of the telescope if the telescope is parallel to the frame. Adjustment for nonparallelism is made to the collar, by means of the two screws provided for this purpose.

Determination of any of the errors should be based upon a series of observations, rather than a single one. This is particularly true in the case of index error, which should be determined by approaching coincidence from opposite directions (up and down) on alternate readings. If adjustments are made carefully, and the sextant is given proper handling, it should remain in adjustment over a long period of time. Unless the navigator has reason to question the accuracy of the adjustments, they need not be checked at intervals of less than several months, except in the case of index error, which has the greatest effect on accuracy of readings, and should be checked each time the sextant is used. If the horizon is used for determining index error, this check should be made *before* evening twilight observations, and *after* morning twilight observations, while the horizon is sharp and distinct. If a star is used, the index error should be determined *after* evening observations and *before* morning sights are taken. During the day, it should be checked both before and after observations.

Frequent manipulation of the adjusting screws should be avoided, as it may cause excessive wear. Except in the case of index error, slight lack of adjustment has little effect on the results, and should be ignored. If adjustments are needed at frequent intervals, the sextant is not receiving proper care, or has worn parts which should be replaced at a navigation instrument shop. If index error is not constant, it should not be removed, but index correction should be determined before or after every observation and applied to the readings, until the sextant can be repaired. A small variable error might well be accepted, but should be watched to see that it does not become unduly large.

Selection of a Sextant

Selection of a sextant—For satisfactory results a sextant should be selected carefully. (See appendix B) For accurate work the radius of the arc should be about 7½ inches or more. The instrument should be light, but strongly built. The various moving parts should fit snugly, but move freely without binding or gritting. If the index arm is either too loose or too tight at either end of the arc, the pivot may not be perpendicular to the frame of the sextant. The telescope should be easy to insert or remove from its holder, and to focus.

The use to be made of a sextant should be considered in its selection. For ordinary use in measuring altitudes of celestial bodies, an arc of 90° or slightly more is sufficient. A longer arc is desirable if back sights are to be made, or if horizontal angles are to be measured. If use of the sextant is to be limited to horizontal angles, less accuracy is required. The arc can be of smaller radius, and small nonadjustable errors are unimportant.

If practicable, a sextant should be examined by an expert, and tested for nonadjustable errors before acceptance.

Octants, quintants, and quadrants—Originally, the term "sextant" was applied to the navigator's double-reflecting, altitude-measuring instrument only if its arc was 60° in length—a sixth of a circle—permitting measurement of angles from 0° to 120°. In modern usage the term is applied to all navigational altitude-measuring instruments, regardless of angular range or principles of operation, although some are octants (angular range 90°), some quintants (144°), some quadrants (180°), and many have an intermediate range.

The artificial horizon—Measurement of altitude requires a horizontal reference. In the case of the marine sextant this is commonly provided by the visible sea horizon. If this is not clearly visible, reliable altitudes cannot be measured unless a different horizontal reference is available. Such a reference is commonly called an **artificial horizon.** If it is attached to, or part of, the sextant, altitudes can be measured at sea, on land, or in the air, whenever celestial bodies are available for observations. On land, where the visible horizon is not a reliable indication of the horizontal, an external artificial horizon can be devised.

Any horizontal reflecting surface will serve the purpose. A pan of mercury, heavy oil, molasses, or other viscous liquid sheltered from the wind is perhaps simplest. A piece of plate glass fitting snugly across the top of the container is usually the best shelter. If there is any reasonable doubt as to the parallelism of the two sides of the glass, two readings should be made with the glass turned 180° in azimuth between readings, and the average value taken. The pan and liquid should be clean, as foreign material on the surface of the liquid is likely to distort the image and introduce an error in the reading.

To use an external artificial horizon, the observer stands or sits in such a position that the celestial body to be observed is reflected in the liquid, and is also visible by direct view. By means of the sextant, the double-reflected image is brought into coincidence with the image appearing in the liquid. In the case of the sun or moon the *bottom* of the double-reflected image is brought into coincidence with the *top* of the image in the liquid, if a lower-limb observation is desired. For an upper-limb observation, the opposite sides are brought into coincidence. If one image is made to cover the other, the observation is of the *center* of the body.

When the observation has been made, apply the index correction and any other instrumental correction, as well as any correction for personal error. Then take *half* the remaining angle and apply all other corrections except dip (height of eye) correction, since this is not applicable. If the *center* of the sun or moon is observed, omit, also, the correction for semidiameter.

A commercial artificial horizon consisting of a metal tray, mercury, cover of two sloping glass sides held in a metal frame, metal bottle to hold the mercury when not in use, and a funnel for pouring, was at one time a familiar part of a navigator's equipment, but the modern navigator might experience difficulty in locating such a device.

Artificial-horizon sextants—Shortly after the marine sextant was invented, an attempt was made to extend its use to periods of darkness. This was done by providing a spirit level attachment. The observer brought the double-reflected image of the celestial body being observed into coincidence with the bubble of the spirit level. Such devices have been made available from time to time, and are still being manufactured. However, they have never come into general use, and are of questionable value.

Charles A. Lindbergh's historic solo flight across the North Atlantic in 1927 demonstrated the practicability of long over-water flights. The development of a suitable instrument for observing altitudes of celestial bodies during darkness and when the horizon was obscured by clouds or haze became a virtual requirement. Various forms of artificial horizon have been used, including a bubble, gyroscope, and pendulum. Of these, the bubble has been most widely used. Figure 13 illustrates a modern periscopic sextant permitting observation with only a small tube protruding through the top of the airplane. Figure 14 shows the optical principle of a different type aircraft sextant.

With an artificial horizon of the bubble or pendulum type, considerable skill is needed to make an observation. The image of the horizontal reference (a circle or horizontal line) and the celestial body both appear in the field of view, and both may seem unsteady. An observation is made by matching the two near the center of the field of view. The appearance at coincidence depends upon the

FIGURE 13—*Modern periscopic sextant.*

instrument. Some bubbles appear dark and are placed on a level with the body. Others have a clear center and are placed over the body. One pendulum type has a horizontal line that is customarily placed directly across the body, although a limb observation can be made if desired. Bubbles can be regulated in size, and the instructions provided with the instrument should be followed. In

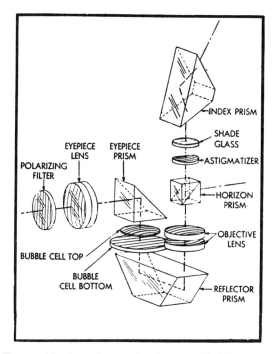

FIGURE 14—*Optical principle of a typical bubble sextant.*

general, the bubble diameter should be about one-sixth to one-fourth the size of the field of view. This is about three to four times the size of the sun or full moon as seen through the eyepiece. A very small bubble should be avoided because it tends to lag sextant movements so much that it is unreliable as a horizontal reference.

A considerable amount of practice is needed to develop skill in making reliable observations with an artificial-horizon sextant, even on land or other steady platform. At sea or in the air the motions of the craft greatly increase the difficulty of observation. In addition to compounding the difficulty of making coincidence, the craft motion introduces a sometimes large and rapidly varying **accelera-**

tion error. That is, motions of the craft produce an acceleration on the pendulum or the liquid of the bubble chamber, causing false indication of the horizontal. In smooth air the accelerations tend to follow a cycle of about one to two minutes in length. They are largely eliminated by use of an averaging device.

In making an observation, the observer attempts to maintain coincidence continuously over a period, usually two minutes. The *average* altitude, generally indicated on a dial or drum, is used with the *mid* time of observation. Thus, perhaps 60 individual observations, or a continuously integrated altitude, are available to smooth out errors of individual observations.

On land or other steady platform a skillful observer using a two-minute averaging bubble or pendulum sextant can measure altitudes to an accuracy of perhaps 2′ (two miles). This, of course, refers to the accuracy of measurement only, and does not include additional errors such as abnormal refraction, deflection of the vertical, computational and plotting errors, etc. In steady flight through smooth air the error of a two-minute observation is increased to perhaps five to ten miles. At sea, conditions are different. In a glassy sea with virtually no roll or pitch, results should approach those on land. However, with even a slight, gentle roll the accelerations to which a vessel is subjected are quite complex, as indicated by the difficulty one not accustomed to the sea has in getting his "sea legs" during the early part of a voyage. If the vessel is yawing, a large error may be introduced. Under these conditions observational errors of 10-15 miles are not unreasonable. With a moderate sea, errors of 30 miles or more are common. In a heavy sea, any useful observations are virtually impossible to obtain. Single altitude observations in a moderate sea can be in error by a matter of *degrees.*

Because of the difficulty of observing, and the large acceleration errors encountered aboard a vessel, bubble and pendulum type sextants have very limited use at sea. A submarine on war patrol, surfacing only during darkness, might have use for such an instrument, but it is impractical for yachts. A large number of observations on a reasonably calm night can produce results of some value. However, even under these conditions some navigators report better results with a marine sextant and dark-adapted eyes. In pack ice a ship generally provides a reasonably steady platform. When the horizon is obscured by ice or haze, polar navigators can sometimes obtain better results with an artificial-horizon sextant than with a marine sextant. Some artificial-horizon sextants have provision for making observations with the natural horizon as a reference, but since this is a secondary usage, results are not generally as satisfactory as by marine sextant. Because of their more complicated optical systems, and the need for providing a horizontal reference, artificial-horizon sextants are generally much more costly to manufacture than marine sextants. Designed for use in the air, they serve a useful purpose there, but for ordinary use aboard ship they have little to recommend them.

Altitudes observed by artificial-horizon sextant are subject to the same errors as those observed by marine sextant, except that dip (height of eye) correction does not apply. Also, when the center of the sun or moon is observed, no correction for semidiameter should be made.

Adjustment of an artificial-horizon sextant should not be attempted by other than an instrument man qualified to handle the particular type instrument involved. An exception is the adjustment of the size of the bubble. Also, with some instruments an easily movable index permits elimination or reduction of index error. This error can best be determined in an instrument shop equipped with a collimator. If one is not available, the error can be determined by comparing the average of a number of observations made at a known point on land with the computed values. A precomputed curve of altitude versus time is useful for this purpose. Altitude corrections equal to the errors but with reversed sign should be applied to computed altitudes. With normal usage, the index error should not change. In most artificial-horizon sextants there is no index error.

The care and operation of various types of instruments vary considerably. The instruction booklet provided with each instrument should serve as the guide. Information on certain artificial-horizon sextants, and a general guide to artificial-horizon sextant observation, is given in H.O. Pub. No. 216, *Air Navigation,* and other texts.

The Marine Chronometer

The marine chronometer is a timepiece having a nearly constant rate. It is used aboard ship to provide accurate time, primarily for timing celestial observations for lines of position, and secondarily for setting the ship's other timepieces. It differs from a watch principally in that it contains a variable lever device to maintain even pressure on the mainspring, and a· special balance designed to compensate for temperature variations. A ship in which celestial navigation is used carries one or more chronometers.

A chronometer is set approximately to Greenwich mean time (GMT) and is not reset until the instrument is overhauled and cleaned, usually at three-year intervals. Resetting might disturb the rate. Instead, the difference between GMT and **chronometer time** (**C**) is carefully determined, and applied as a correction to all chronometer readings. This difference, called **chronometer error** (**CE**), is "fast" (**F**) if chronometer time is later than GMT, and "slow" (**S**) if earlier. The amount by which chronometer error changes in one day is called **chronometer rate,** or sometimes **daily rate,** considered "gaining" or "losing" as the chronometer is running faster or slower than the correct rate. An erratic rate indicates a defective instrument, or need for overhaul.

A chronometer is mounted in gimbals in a box, which should be carefully stowed to protect the instrument from damage due to heavy rolling and pitching, vibration, temperature variations, and electrical and magnetic influences. Usually this is done by fitting the box snugly into a heavily padded case suitably located in the chart room of

43

merchant ships, and below decks, near the center of motion, in U. S. Navy ships.

The principal maintenance requirement aboard ship is regular winding at about the same time each day. Aboard United States naval vessels this is customarily done at about 1130 each morning, and reported to the commanding officer at 1200. Aboard merchant ships it is usually wound at about 0800. Although a chronometer is designed to run for more than two days, daily winding helps insure a uniform rate, and constitutes a daily routine that decreases the possibility of letting the instrument run down. On the face of each chronometer is a small dial that indicates the number of hours before the chronometer will be run down. To wind the chronometer, gently turn the instrument on its side, and slide back the guard covering the keyhole. Insert the key and carefully wind in a counterclockwise direction. Seven half-turns should suffice. If a chronometer should run down, wait until GMT is nearly the same as the time indicated before winding. If the chronometer does not start after winding, move the case back and forth gently. Check the error and rate carefully.

At maximum intervals of about three years, a chronometer should be sent to a good chronometer repair shop for cleaning and overhaul. When transported by hand, a chronometer should be clamped in its gimbals and stowed in the large case provided. When shipped, it should be allowed to run down, and the balance secured by a cork before the chronometer is stored in the large case.

Watches

Watches—In the interest of accuracy, a chronometer is not disturbed more than necessary. Celestial observations are timed and ship's clocks set by means of a **comparing watch.** This is a high-grade pocket watch which is set by comparison with a chronometer, and then carried to the place where accurate time is needed. For celestial navigation, a comparing watch should have a large sweep-second hand which can be set. A comparing watch used for timing celestial observations should preferably be set to Greenwich mean time, to avoid the necessity of applying a correction for each observation.

If the second hand cannot be set, the watch should be set to the nearest whole minute, being sure that the second hand is in synchronism with the minute hand, and the **watch error (WE)** determined. If a watch is to be used for other purposes than timing of celestial observations, it might preferably be set to zone time. A comparing watch should be set, or watch error determined, immediately before or after celestial observations are made, to avoid the necessity for determining and applying a correction for **watch rate,** and to eliminate a possible error due to an inaccurate or variable rate. If a watch set to GMT is used for timing celestial observations, care should be taken to avoid a possible error of 12 hours or 24 hours. The mental application of zone description to ship's time indicates the approximate GMT and the Greenwich date. A stop watch may also be used for celestial observations.

Other instruments—The sextant, chronometer, and comparing watch (or stop watch) are the principal instruments of celestial navigation. Plotting equipment is the same as that for dead reckoning. A flashlight might be needed for reading the sextant and the comparing watch. A pocket notebook is desirable for recording predicted positions of celestial bodies if a star finder is used, and for recording the observations. A workbook is desirable for solving celestial observations, so that a permanent record is available. Work forms are desirable, but should form part of the work book, and not be kept separately. These might be provided by rubber stamp, or by printing. In the latter case a looseleaf work book may be desirable to permit arrangement of the various papers in chronological order.

IDENTIFICATION OF
CELESTIAL BODIES

Sky Charts

A basic requirement of celestial navigation is the ability to identify the bodies observed. This is not difficult because relatively few celestial bodies are commonly used for navigation, and various aids are available to assist in their identification, as explained in this chapter.

Many navigators consider it a matter of professional pride to have a more extensive acquaintance with the heavens than required by the relatively simple demands of navigation.

Bodies of the solar system—No problem is encountered in the identification of the sun and moon. However, the planets can be mistaken for stars. A person working continually with the night sky recognizes a planet by its changing position among the relatively fixed stars. He identifies the planets by noting their positions relative to each other, the sun, the moon, and the stars. He knows that they remain within the narrow limits of the zodiac, but are in almost constant motion relative to the stars. The magnitude and color may be helpful. The information he needs is found in the *Nautical Almanac*. The "Planet

Notes" near the front of that volume are particularly useful.

Sometimes the light from a planet seems steadier than that from a star. This is because fluctuation of the unsteady atmosphere causes **scintillation** or **twinkling** of a star, which has no measurable diameter with even the most powerful telescopes. The navigational planets are less susceptible to twinkling because of the broader apparent area giving light.

Planets can also be identified by the *Air Almanac* ecliptic diagram, star finder, sky diagram, or by computation, each of which is discussed later.

Stars—The average navigator regularly uses not more than perhaps 20 or 30 stars. The *Nautical Almanac* gives full navigational information on 19 first magnitude stars and 38 second magnitude stars, in addition to Polaris. Abbreviated information is given for 115 more. Additional stars are listed in *The American Ephemeris and Nautical Almanac* and in various star catalogs. About 6,000 stars of the sixth magnitude or brighter (on the entire celestial sphere) are visible to the unaided eye on a clear, dark night.

Stars are designated by one or more of the following:

Name: Most names of stars, as now used, were given by the ancient Arabs and some by the Greeks or Romans. One of the stars of the *Nautical Almanac,* Nunki, was named by the Babylonians. Only a relatively few stars have names. Several of the stars on the daily pages of the almanacs had no name prior to the 1953 edition, and were given coined names so that all stars listed on the daily pages might have names. The pronunciation, meaning, and other information of general interest regarding Polaris and the 57 stars listed on the daily pages of the *Nautical Almanac* are given in appendix A.

Bayer's name: Most bright stars, including those with names, have been given a designation consisting of a Greek letter followed by the possessive form of the name of the constellation, as α *Cygni* (Deneb, the brightest star

48

in the constellation *Cygnus,* the swan). Roman letters are used when there are not enough Greek letters. Usually, the letters are assigned in order of brightness within the constellation, but in some cases the letters are assigned in another order, where it seems logical to do so. An example is the big dipper, where the letters are assigned in order from the outer rim of the bowl to the end of the handle. This system of star designation was suggested by John Bayer of Augsburg, Germany, in 1603. All of the 173 stars included in the list near the back of the *Nautical Almanac* are given by Bayer's name as well as regular name, where there is one.

Flamsteed's number: A similar system, accommodating more stars, numbers them in each constellation, from west to east, the order in which they cross the celestial meridian. An example is 95 *Leonis,* the 95th star in the constellation *Leo,* the lion. This system was suggested by John Flamsteed (1646-1719), who was the first British Astronomer Royal.

Catalog number: Stars are sometimes designated by the name of a star catalog and the number of the star as given in that catalog, as A. G. Washington 632. In these catalogs stars are listed in order from west to east, without regard to constellation, starting with the hour circle of the vernal equinox. This system is used primarily for dimmer stars having no other designation. Navigators seldom have occasion to use this system.

The ability to identify stars by position relative to each other is useful to the navigator. A tabulation of the relative positions of the 57 stars given on the daily pages of the *Nautical Almanac,* and Polaris, is given in appendix G. A star chart is helpful in locating these relationships and others which may be useful. This method is limited to periods of relatively clear, dark skies with little or no overcast. Stars can also be identified by the *Air Almanac* ecliptic diagram, star finder, H.O. Pub. No. 249, sky diagram, or by computation.

Star charts are based upon the celestial equator system

of coordinates, using declination and sidereal hour angle (or right ascension). The zenith of the observer is at the intersection of the parallel of declination equal to his latitude, and the hour circle coinciding with his celestial meridian. This hour circle has an SHA equal to $360° -$ LHAϒ (or RA = LHAϒ). The horizon is everywhere $90°$ from the zenith. A **star globe** is similar to a terrestrial sphere, but with stars (and often constellations) shown instead of geographical positions. Star globes are used by British navigators, but not customarily by Americans. The combined *Nautical Almanac* includes instructions for using this device. On a star globe the celestial sphere is shown as it would appear to an observer *outside* the sphere. Constellations appear reversed. Star charts may show a similar view, but more often they are based upon the view from *inside* the sphere, as seen from the earth. On these charts, north is at the top, as with maps, but east is to the *left* and west to the *right*. The directions seem correct when the chart is held overhead, with the top toward the north, so that the relationship is similar to that in the sky. Any map projection can be used, but some are more suitable than others.

The *Nautical Almanac* has four star charts. The two principal ones are on the polar azimuthal equidistant projection, one centered on each celestial pole. Each chart extends from its pole to declination $10°$ (same name as pole). Below each polar chart is an auxiliary chart on the Mercator projection, from $30°$ N to $30°$ S. On any of these charts, the zenith can be located as indicated above, to determine which stars are overhead. The horizon is $90°$ from the zenith. The charts can also be used to determine the location of a star relative to surrounding stars. The *Air Almanac* has a fold-in chart at the back, on the rectangular projection. This projection is suitable for indicating the coordinates of the stars, but excessive distortion occurs in regions of high declination. The celestial poles are represented by the top and bottom horizontal *lines* the same length as the celestial equator. To locate the horizon on this chart, first locate the zenith as indicated above, and then locate the four cardinal points. The north and south points are 90° from the zenith, along the celestial meridian. The distance to the elevated pole (having the same name as the latitude) is equal to the colatitude of the observer. The remainder of the 90° (the latitude) is measured *from* the same pole, along the *lower branch* of the celestial meridian, 180° from the upper branch containing the zenith. The east and west points are on the celestial equator at the hour circle 90° east and west (or 90° and 270° in the same direction) from the celestial meridian. The horizon is a sine curve through the four cardinal points. Directions on this projection are distorted.

The star charts shown in figures 15-18, on the transverse Mercator projection, are designed to assist one in learning the stars listed on the daily pages of the *Nautical Almanac,* and Polaris. Each chart extends about 20° beyond each celestial pole, and about 60° (four hours) each side of the central hour circle (at the celestial equator). Therefore, they do not coincide exactly with that half of the celestial sphere above the horizon at any one time or place. The zenith, and hence the horizon, varies with the position of the observer on the earth, and also with the rotation of the earth (apparent rotation of the celestial sphere). The charts show all stars of fifth magnitude and brighter as they appear in the sky, but with some distortion toward the right and left edges.

Only Polaris and the 57 stars listed on the daily pages of the *Nautical Almanac* are named on the charts. The almanac star charts should be used for locating the additional stars given near the back of the *Nautical Almanac.* The broken lines connect stars of some of the more prominent constellations. The solid lines indicate the celestial equator and certain useful relationships among stars in different constellations. The celestial poles are marked by crosses, and labeled. By means of the celestial equator and the poles, one can locate his zenith approximately along the mid hour circle, when this coincides with his celestial meridian, as shown in the table below. At any time earlier than those shown in the table the zenith is to the *right* of center, and at a later time it is to the *left,* approximately one-quarter of the distance from the center to the outer edge (at the celestial equator) for each hour that the time differs from that shown. The stars in the vicinity of the north pole can be seen in proper perspective by inverting the chart, so that the zenith of an observer in the northern hemisphere is *up* from the pole.

	Fig. 15	*Fig. 16*	*Fig. 17*	*Fig. 18*
Local sidereal time	0000	0600	1200	1800
LMT 1800	Dec. 21	Mar. 22	June 22	Sept. 21
LMT 2000	Nov. 21	Feb. 20	May 22	Aug. 21
LMT 2200	Oct. 21	Jan. 20	Apr. 22	July 22
LMT 0000	Sept. 22	Dec. 22	Mar. 23	June 22
LMT 0200	Aug. 22	Nov. 22	Feb. 21	May 23
LMT 0400	July 23	Oct. 22	Jan. 21	Apr. 22
LMT 0600	June 22	Sept. 21	Dec. 22	Mar. 23

Stars in the vicinity of *Pegasus* (fig. 15)—In autumn the evening sky has few first magnitude stars. Most of these are near the southern horizon of an observer in the latitudes of the United States. A relatively large number of second and third magnitude stars seem conspicuous, perhaps because of the small number of brighter stars. High in the southern sky three third magnitude stars and one second magnitude star form a square with sides nearly 15° of arc in length. This is *Pegasus,* the winged horse, although to many modern men it more nearly resembles a baseball diamond, complete with catcher, pitcher, batter, umpire, base umpire near second base, infield and out-field; although there does seem to be a large number of outfielders. One may even see the next batter, bat boy, and coach.

Only Markab at the southwestern corner (third base) and Alpheratz at the northeastern corner (first base) are

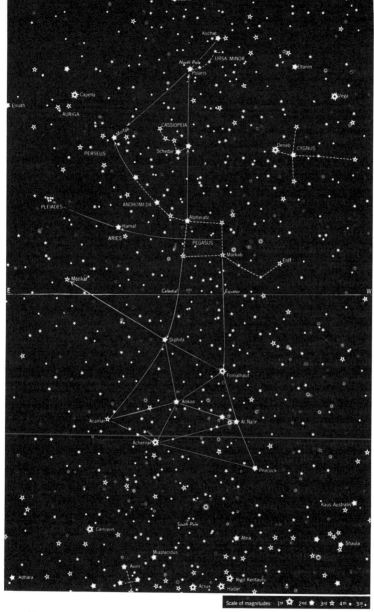

FIGURE 15—*Stars in the vicinity of* Pegasus.

listed on the daily pages of the *Nautical Almanac*. Alpheratz is part of the constellation *Andromeda*, the princess, extending in an arc toward the northeast and terminating at Mirfak in *Perseus*, legendary rescuer of *Andromeda*.

A line extending northward through the eastern side (first-second base line) of the square of *Pegasus* passes through the leading (western) star of M-shaped (or W-shaped) *Cassiopeia*, the legendary mother of the princess *Andromeda*. The only star of this constellation listed on the daily pages of the *Nautical Almanac* is Schedar, the second star from the leading one as the configuration circles the pole in a counterclockwise direction. If the line through the eastern side of the square of *Pegasus* is continued on toward the north, it leads to second magnitude Polaris, the north star (less than 1° from the north celestial pole) and brightest star of *Ursa Minor*, the little bear. Kochab, a second magnitude star at the other end of the little dipper, is also listed in the almanacs. At this season the big dipper is low in the northern sky, below the celestial pole. A line extending from Kochab through Polaris leads to Mirfak, assisting in its identification when *Pegasus* and *Andromeda* are near or below the horizon.

Deneb, in *Cygnus*, the swan, and Vega are bright, first magnitude stars in the northwestern sky. They are discussed in article 2208. Capella, a bright star in the northeastern sky, is discussed later.

The line through the eastern side of the square of *Pegasus* (first-second base line) approximates the hour circle of the vernal equinox, shown at ♈ on the celestial equator to the south. The sun is at ♈ on or about March 21, when it crosses the celestial equator from south to north. If the line through the eastern side of *Pegasus* is extended southward and curved slightly toward the east, it leads to second magnitude Diphda. A longer and straighter line southward through the western side (home plate-third base line) of *Pegasus* leads to first magnitude Fomalhaut. A line extending northeasterly from Fomalhaut through Diphda leads to Menkar, a third magnitude star, but the brightest in its vicinity. Ankaa, Diphda, and Fomalhaut form an isosceles triangle, with the apex at Diphda. Ankaa is near or below the southern horizon of observers in latitudes of the United States. Four stars farther south than Ankaa may be visible when on the celestial meridian, just above the horizon of observers in latitudes of the extreme southern part of the United States. These are Acamar, Achernar, Al Na'ir, and Peacock. These stars, with each other and with Ankaa, Fomalhaut, and Diphda, form a series of triangles as shown in figure 15. Almanac stars near the bottom of figure 15 are discussed in succeeding paragraphs.

Two other almanac stars can be located by their positions relative to *Pegasus.* These are Hamal in the constellation *Aries,* the ram, east of *Pegasus,* and Enif, west of the southern part of the square, identified as shown in figure 15. The line leading to Hamal, if continued, leads to the *Pleiades,* not used by navigators for celestial observations, but a prominent figure in the sky, heralding the approach of the many conspicuous stars of the winter evening sky, figure 16.

Stars in the vicinity of *Orion* (fig. 16)—As *Pegasus* leaves the meridian and moves into the western sky, *Orion,* the mighty hunter, rises in the east. With the possible exception of the big dipper, no other configuration of stars in the entire sky is as well known as *Orion* and its immediate surroundings. In no other part are there so many first magnitude stars.

The belt of *Orion,* being nearly on the celestial equator, is visible by an observer in virtually any latitude, rising and setting almost on the prime vertical, and dividing equally its time above and below the horizon. Of the three second magnitude stars forming the belt, only Alnilam, the middle one, is listed on the daily pages of the *Nautical Almanac.*

Four conspicuous stars form a box around the belt. To the south is Rigel, one of the hottest and bluest of the stars,

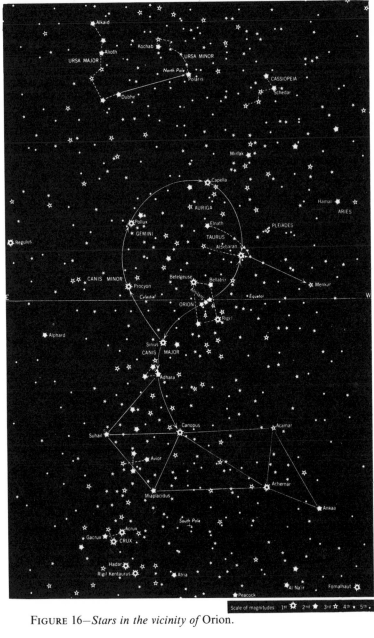

Scale of magnitudes: 1st ☆ 2nd ★ 3rd ☆ 4th ★ 5th .

FIGURE 16—*Stars in the vicinity of* Orion.

in contrast with relatively cool, red, variable Betelgeuse, at approximately an equal distance to the north. Bellatrix, bright for a second magnitude star but overshadowed by its more brilliant neighbors, is a few degrees west of Betelgeuse. Neither the second magnitude star forming the southeastern corner of the box, nor any star of the dagger, is listed on the daily pages of the *Nautical Almanac.*

A line extending eastward from the belt of *Orion* and curving toward the south leads to Sirius, the brightest star in the entire heavens, having a magnitude of (−) 1.6. Only Mars and Jupiter at or near their greatest brilliance, and the sun, moon, and Venus are brighter than Sirius. This is part of the constellation *Canis Major,* the large hunting dog of *Orion.* Starting at Sirius a curved line extends northward through first magnitude Procyon, in *Canis Minor,* the small hunting dog; first magnitude Pollux and second magnitude Castor (not listed on the daily pages of the *Nautical Almanac*), the twins of *Gemini;* brilliant Capella in *Auriga,* the charioteer; and back down to first magnitude Aldebaran, the follower, which trails the *Pleiades,* the seven sisters. Aldebaran, brightest star in the head of *Taurus,* the bull, may also be found by a curved line extending northwestward from the belt of *Orion.* The V-shaped figure forming the outline of the head and horns of *Taurus* points toward third magnitude Menkar. At the summer solstice the sun is between Pollux and Aldebaran.

If the curved line from *Orion's* belt southeastward to Sirius is continued, it leads to a conspicuous, small, nearly equilateral triangle of three bright second magnitude stars of nearly equal brilliancy. This is part of *Canis Major.* Only Adhara, the westernmost of the three stars, is listed on the daily pages of the *Nautical Almanac.* Continuing on with somewhat less curvature, the line leads to Canopus, second brightest star in the heavens and one of the two stars having a negative magnitude (−0.9). With Suhail and Miaplacidus, Canopus forms a large, equilateral triangle which partly encloses the false southern cross. The brightest star within this triangle is Avior, near its center.

Canopus is also at one apex of a triangle formed with Adhara to the north and Suhail to the east, another triangle with Acamar to the west and Achernar to the southwest, and another with Achernar and Miaplacidus. Acamar, Achernar, and Ankaa form still another triangle toward the west. Because of chart distortion, these triangles do not appear in the sky in exactly the relationship shown on the star chart. Other daily-page almanac stars near the bottom of figure 16 are discussed in succeeding articles.

During the winter evening sky the big dipper is east of Polaris, the little dipper is nearly below it, and *Cassiopeia* is west of it. Mirfak is northwest of Capella, nearly midway between it and *Cassiopeia*. Hamal is in the western sky. Regulus and Alphard are low in the eastern sky, heralding the approach of the configurations associated with the evening skies of spring.

Stars in the vicinity of *Ursa Major* (fig. 17)—As if to enhance the splendor of the sky in the vicinity of *Orion,* the region toward the east, like that toward the west, has few bright stars, except in the vicinity of the south celestial pole. However, as *Orion* sets in the west, leaving Capella and Pollux in the northwestern sky, a number of good navigational stars move into favorable positions for observation.

The big dipper, part of *Ursa Major,* the great bear, appears prominently *above* the north celestial pole, directly opposite *Cassiopeia* (only partly shown in fig. 17), which appears as a W just above the northern horizon of most observers in latitudes of the United States. Of the seven stars forming the big dipper, only Dubhe, Alioth, and Alkaid are listed on the daily pages of the *Nautical Almanac.*

The two second magnitude stars forming the outer part of the bowl of the big dipper are often called the *pointers* because a line extending northward (*down* in spring evenings) through them points to Polaris. The little dipper, with Polaris at one end and Kochab at the other, is part of

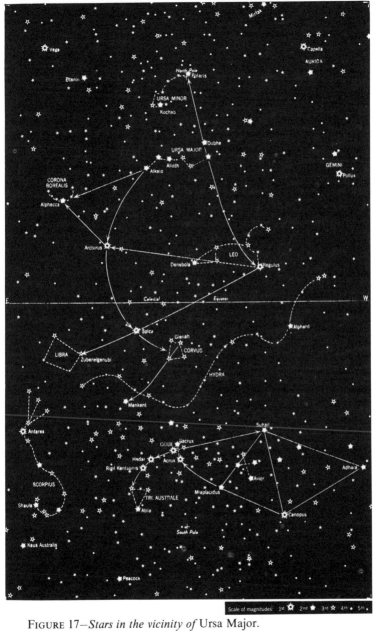

Scale of magnitudes: 1st ✦ 2nd ✦ 3rd ✦ 4th ✦ 5th •

FIGURE 17—*Stars in the vicinity of* Ursa Major.

59

Ursa Minor, the little bear. Relative to its bowl, the handle of the little dipper curves in the opposite direction to that of the big dipper. Other almanac stars near the top of figure 17 are discussed elsewhere.

A line extending southward through the pointers, and curving somewhat toward the west, leads to first magnitude Regulus, brightest star in *Leo,* the lion. The head, shoulders, and front legs of this constellation form a sickle, with Regulus at the end of the handle. Toward the east is second magnitude Denebola, the tail of the lion. On toward the southwest from Regulus is second magnitude Alphard, brightest star in *Hydra,* the sea serpent. A dark sky and considerable imagination are needed to trace the long, winding body of this figure.

A curved line extending the arc of the handle of the big dipper leads to first magnitude Arcturus. With Alkaid and Alphecca, brightest star in *Corona Borealis,* the northern crown, Arcturus forms a large, inconspicuous triangle. If the arc through Arcturus is continued, it leads next to first magnitude Spica and then to *Corvus,* the crow, which appears most like a gaff mainsail of a schooner. The brightest star in this constellation is Gienah, but three others are nearly as bright. At autumnal equinox the sun is on the celestial equator, about midway between Regulus and Spica.

A long, slightly curved line from Regulus east-south-easterly through Spica leads to Zubenelgenubi (zōō.běn′ěl.jě.nū′bē) at the southwestern corner of an inconspicuous box-like figure called *Libra,* the (weighing) scales.

Returning to *Corvus,* a line from Gienah, extending diagonally across the figure and then curving somewhat toward the east, leads to Menkent, just beyond *Hydra.*

Far to the south, below the horizon of most northern-hemisphere observers, a group of bright stars is a prominent feature of the spring sky of the southern hemisphere. *Crux,* the southern cross, is about 40° south of *Corvus.* This is a small figure and a poor cross, and hence disappointing to many who view it for the first time. The "false cross" to the west is a better but less conspicuous cross. Acrux at the southern end of the southern cross, and Gacrux at the northern end, are listed on the daily pages of the *Nautical Almanac.*

The triangles formed by Suhail, Miaplacidus, and Canopus, and by Suhail, Adhara, and Canopus, are west of the southern cross, Suhail being in line with the horizontal arm of the southern cross at this time. A line from Canopus, through Miaplacidus, curved slightly toward the north, leads to Acrux. A line through the east-west arm of *Crux*, eastward and then curving toward the south, leads first to Hadar and then to Rigil Kentaurus, two very bright stars. Continuing on, the curved line leads to small *Triangulum Australe*, the southern triangle, the easternmost star of which is Atria.

Scorpius, the scorpion, Kaus Australis, and Peacock, in the southeastern sky of the southern hemisphere, are discussed in article 2208.

Stars in the vicinity of *Cygnus* (fig. 18)—As the celestial sphere continues in its apparent westward rotation, the stars familiar to a spring evening observer sink low in the western sky. By midsummer the big dipper has moved to a position to the left of the north celestial pole, and the line from the pointers to Polaris is nearly horizontal. The little dipper is standing on its handle, with Kochab above and to the left of the celestial pole. *Cassiopeia* is at the right of Polaris, opposite the handle of the big dipper.

The only first magnitude star in the western sky is Arcturus, which forms a large, inconspicuous triangle with Alkaid, the end of the handle of the big dipper, and Alphecca, the brightest star in *Corona Borealis*, the northern crown.

The eastern sky is dominated by three very bright stars. The westernmost of these is Vega, the brightest star north of the celestial equator, and third brightest star in the heavens. Its magnitude is 0.1. Having a declination of a little less than 39° N, this star passes through the zenith

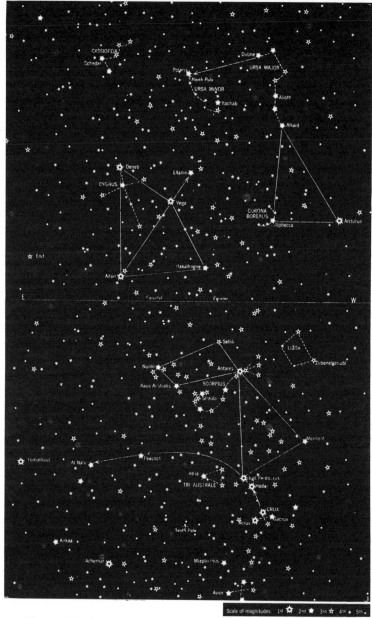

FIGURE 18—*Stars in the vicinity of* Cygnus.

along a path across the central part of the United States, from Washington in the east to San Francisco on the Pacific coast. Vega forms a large but conspicuous triangle with its two bright neighbors, Deneb to the northeast and Altair to the southeast. The angle at Vega is nearly a right angle. Deneb is at the end of the tail of *Cygnus,* the swan. This configuration is sometimes called the northern cross, with Deneb at the head. To modern youth it more nearly resembles a dive bomber while it is still well toward the east, with Deneb at the nose of the fuselage. Altair has two fainter stars close by, on opposite sides. The line formed by Altair and its two fainter companions, if extended in a northwesterly direction, passes through Vega, and on to second magnitude Eltanin. The angular distance from Vega to Eltanin is about half that from Altair to Vega. Vega and Altair, with second magnitude Rasalhague to the west, form a large equilateral triangle. This is less conspicuous than the Vega-Deneb-Altair triangle because the brilliance of Rasalhague is much less than that of the three first magnitude stars, and the triangle is overshadowed by the brighter one.

Far to the south of Rasalhague, and a little toward the west, is a striking configuration called *Scorpius,* the scorpion. The brightest star, forming the head, is red Antares. At the tail is Shaula.

Antares is at the southwestern corner of an approximate parallelogram formed by Antares, Sabik, Nunki, and Kaus Australis. With the exception of Antares, these stars are only slightly brighter than a number of others nearby, and so this parallelogram is not a striking figure. At winter solstice the sun is a short distance northwest of Nunki.

Northwest of *Scorpius* is the box-like *Libra,* the (weighing) scales, in which Zubenelgenubi marks the southwest corner.

With Menkent and Rigil Kentaurus to the southwest, Antares forms a large but unimpressive triangle. For most observers in the latitudes of the United States, Antares is low in the southern sky, and the other two stars of the triangle are below the horizon. To an observer in the southern hemisphere *Crux,* the southern cross, is to the right of the south celestial pole, which is not marked by a conspicuous star. A long, curved line starting with the now-vertical arm of the southern cross and extending northward and then eastward passes successively through Hadar, Rigil Kentaurus, Peacock, and Al Na'ir.

Fomalhaut is low in the southeastern sky of the southern hemisphere observer, and Enif is low in the eastern sky at nearly any latitude. With the appearance of these stars it is not long before *Pegasus* will appear over the eastern horizon during the evening, and as the winged horse climbs evening by evening to a position higher in the sky, a new annual cycle approaches.

Ecliptic diagram—On each right-hand page of the daily tabulations of the *Air Almanac* an ecliptic diagram shows a band of the sky 16° wide (the zodiac) with the sun at the center. Shown in correct position relative to the sun (except when very close to it) are the moon, selected planets and stars, and the vernal equinox. This diagram is useful for planning purposes and for locating the planets. That part of the diagram to the left of the sun is east of it, approximately coinciding with the visible part during evening twilight. That part to the right, or west, of the sun coincides approximately with the visible portion during morning twilight. The two ends are that point in the sky 180° from the sun.

Star Finders

Star finders—Various devices have been invented to help an observer locate and identify individual stars. The most widely used is the Star Finder and Identifier published by the U. S. Navy Hydrographic Office. The current model, H.O. 2102-D, as well as the previous 2102-C model patented by E. B. Collins, formerly of that Office, employs the same basic principle as that used in the Rude Star Finder, which was patented by Captain G. T. Rude and later sold to the Hydrographic Office. Successive models reflect various modifications to meet changing conditions and requirements.

The *star base* of H.O. 2102-D consists of a thin, white, opaque, plastic disk about 8½ inches in diameter, with a small peg in the center. On one side the north celestial pole is shown at the center, and on the opposite side the south celestial pole is at the center. All of the stars listed on the daily pages of the *Nautical Almanac* are shown on a polar azimuthal equidistant projection extending to the opposite pole. The south pole side is shown in figure 19. Many copies of an older edition, H.O. 2102-C, showing the stars listed in the almanacs prior to 1953, and having other minor differences, are still in use. These are not rendered obsolete by the newer edition, but should be corrected by means of the current almanac. The rim of each side is graduated to half a degree of LHA♈ (or 360°−SHA).

Ten transparent *templates* of the same diameter as the star base are provided. There is one template for each 10° of latitude, labeled 5°, 15°, 25°, etc., plus a tenth (printed in red) showing meridian angle and declination. The older edition (H.O. 2102-C) did not have the red meridian angle-declination template. Each template can be used on

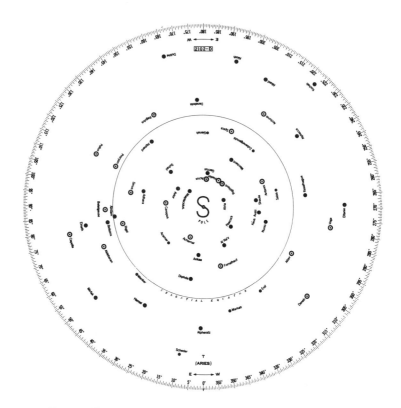

FIGURE 19—*The south pole side of the star base of H.O. 2102-D.*

either side of the star base, being centered by placing a small center hole in the template over the center peg of the star base. Each latitude template has a family of altitude curves at 5° intervals from the horizon (from altitude 10° on the older H.O. 2102-C) to 80°. A second family of curves, also at 5° intervals, indicates azimuth. The north-south azimuth line is the celestial meridian. The star base, templates, and a set of instructions are housed in a circular leatherette container.

Since the sun, moon, and planets continually change

apparent position relative to the "fixed" stars, they are not shown on the star base. However, their positions at any time, as well as the positions of additional stars, can be plotted. To do this, determine 360°−SHA of the body. For the stars and planets, SHA is listed in the *Nautical Almanac.* For the sun and moon, 360°−SHA is found by subtracting GHA of the body from GHAϒ at the same time. Locate 360°−SHA on the scale around the rim of the star base. A straight line from this point to the center represents the hour circle of the body. From the celestial equator, shown as a circle midway between the center and the outer edge, measure the declination (from the almanac) of the body *toward* the center if the pole and declination have the *same* name *(both* N or *both* S), and *away* from the center if they are of *contrary* name. Use the scale along the north-south azimuth line of any template as a declination scale. The meridian angle-declination template (the latitude 5° template of H.O. 2102-C) has an open slot with declination graduations along one side, to assist in plotting positions, as shown in figure 20. In the illustration the celestial body being located has a 360°− SHA of 285°, and a declination of 14°.5 S. It is not practical to attempt to plot to greater precision than the nearest 0°.1. Positions of Venus, Mars, Jupiter, and Saturn on June 1, 1958, are shown plotted on the star base in figure 21. It is sometimes desirable to plot positions of the sun and moon, to assist in planning. Plotted positions of stars need not be changed. Plotted positions of bodies of the solar system should be replotted from time to time, the more rapidly moving ones oftener than others. The satisfactory interval for each body can be determined by experience. It is good practice to record the date of each plotted position of a body of the solar system, to serve later as an indication of the interval since it was plotted.

To orient the template properly for any given time, proceed as follows: enter the almanac with GMT, and determine GHAϒ at this time. Apply the longitude to GHAϒ, subtracting if west or adding if east, to determine

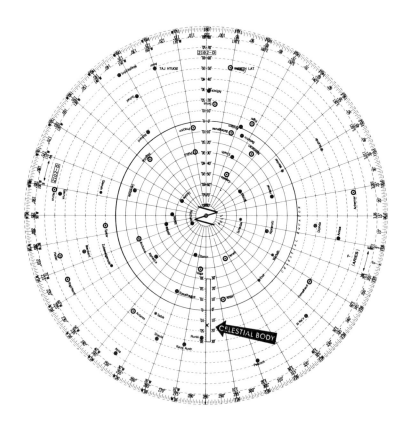

FIGURE 20—*Plotting a celestial body on the star base of H.O. 2102-D.*

LHAȢ. If LMT is substituted for GMT in entering the almanac, LHAȢ can be taken directly from the almanac,

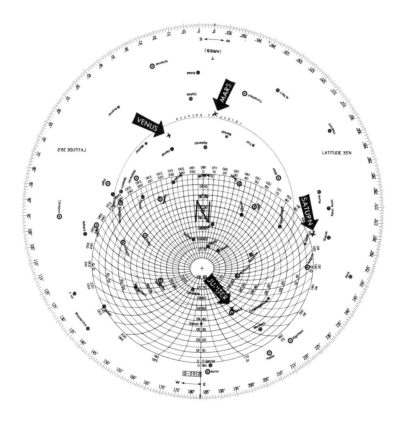

FIGURE 21—*A template in place over the star base of H.O. 2102-D.*

to sufficient accuracy for orienting the star finder template. Select the template for the latitude nearest that of the

69

observer, and center it over the star base, being careful that the correct sides (north or south to agree with the latitude) of both template and star base are used. Rotate the template relative to the star base until the arrow on the celestial meridian (the north-south azimuth line) is over LHAϒ on the star base graduations. The small cross at the origin of both families of curves now represents the zenith of the observer. The approximate altitude and azimuth of the celestial bodies above the horizon can be read directly from the star finder, using eye interpolation. Consider Polaris, not shown, as at the north celestial pole. For more accurate results, the template can be lifted clear of the center peg of the star base, and shifted along the celestial meridian until the latitude, on the altitude scale, is over the pole. This refinement is not needed for normal use of the device. It should not be used for a latitude differing more than 5° from that for which the curves were drawn. If the altitude and azimuth of an identified body shown on the star base are known, the template can be oriented by rotating it until it is in correct position relative to that body.

Customarily, H.O. 2102-D is used in either of two ways:

1. To make an advance list of celestial bodies available for observation at a given time.

2. To identify an unknown celestial body which has been observed.

Example 1—During evening twilight on June 1, 1958, the GMT 2324 DR position of a ship is lat. 34°12′5 N, long. 57°46′8 W.

Required—The approximate altitude (h*a*) and azimuth of each first magnitude star, and any planets, between altitudes 15° and 75°.

Solution (fig. 21)—(1) Plot the positions of the planets, as shown. The values used are those for GMT 1200 on June 1, as follows:

Planet	360° − SHA	Dec.
Venus	28°.5	9°.6 N
Mars	356°.7	3°.5 S
Jupiter	201°.1	7°.3 S
Saturn	262°.8	21°.8 S

(2) Determine LHA♈ by means of the *Nautical Almanac*, as follows:

GMT	2324	June 1
23ʰ	234°55'.0	
24ᵐ	6°01'.0	
GHA♈	240°56'.0	
λ	57°46'.8	W
LHA♈	183°09'.2	

(3) Select the template for latitude 35°, place it over the north side of the star base with "LATITUDE 35° N" appearing correctly, and orient it to 183°.2. It is customary to list the bodies in order of increasing azimuth, as follows:

Body	ha	Zn
Vega	17°	054°
Arcturus	59°	111°
Jupiter	45°	155°
Spica	42°	157°
Regulus	53°	240°
Procyon	20°	262°
Pollux	33°	284°
Capella	15°	316°

Example 2—At the time and place of example 1, an unidentified celestial body is observed through a break in the clouds. Its sextant altitude is 15°27'.8, and its azimuth is 085°.

Required—Identify the celestial body.

Solution (fig. 21)—Orient the template as in Example 1. By means of its altitude and azimuth, identify the star as Rasalhague.

71

If no body appears at the measured altitude and azimuth, place the red meridian angle-declination template over the altitude-azimuth template and read off, by inspection, the declination and the 360°—SHA value of the body, and from this, determine its SHA. Using the SHA and declination, enter the list of stars near the back of the *Nautical Almanac,* and identify the body. If it is not found in this list, and no error has been made, one of the stars not listed in the almanac, or possibly the planet Mercury, has been observed. Unless a copy of *The American Ephemeris and Nautical Almanac* or another book containing the required information is available, the observation cannot be used. If right ascension (measuring *eastward* in time units, through 24 hours) of the body is available, but not its SHA, the value taken from the star finder (360°—SHA) is converted to time units (art. 1904) and used directly, since RA = 360°—SHA.

Example 3—At the time and place of Example 1 an unidentified celestial body is observed through a break in the clouds. Its altitude is 52°58′9, and its azimuth is 170°.

Required—Identify the celestial body.

Solution (fig. 2210c)—Orient the template as in example 1. Since no celestial body appears at the place indicated by its altitude and azimuth, the red meridian angle-declination template is placed over the altitude-azimuth template. The declination is found to be about 1° S. The 360°—SHA value is about 190°, and SHA is therefore about 170° From the star list near the back of the *Nautical Almanac,* the star is identified as γ *Virginis.*

Sight Reduction Tables for Air Navigation (H.O. Pub. No. 249)—Volume I of H.O. Pub. No. 249 can be used as a star finder for the stars tabulated at any given time. For these bodies the altitude and azimuth are tabulated for each 1° of latitude and 1° of LHAϒ (2° beyond latitude 69°). The principal limitation is the small number of stars listed.

Sky diagram—Near the back of the *Air Almanac* are a number of **sky diagrams.** These are azimuthal equidistant

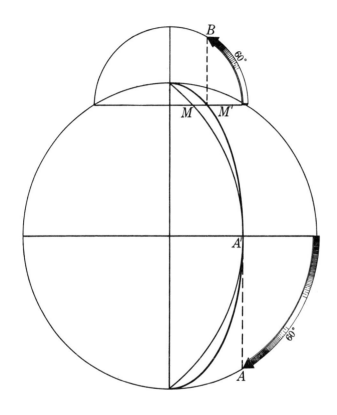

FIGURE 22—*Locating a point on an ellipse of a diagram on the plane of the celestial meridian.*

projections of the celestial sphere on the plane of the horizon, at latitudes 70°N, 50°N, 30°N, 10°N, 10°S, and 30°S, at intervals of two hours of local mean time each month. A number of the brighter stars, the visible planets, and several positions of the moon are shown at their correct altitude and azimuth. These are of limited value because of their small scale; the large increments of latitude, time, and date; and the limited number of bodies shown. However, in the absence of other methods, particularly a star finder, these diagrams can be useful. Allowance can be made for variations from the conditions for which each diagram is constructed. Instructions for use of the diagrams are included in the *Air Almanac.*

Identification by computation—If the altitude and azimuth of the celestial body, and the approximate latitude of the observer, are known, the navigational triangle can be solved for meridian angle and declination. The meridian angle can be converted to LHA, and this to GHA. With this and GHA of Aries at the time of observation, the SHA of the body can be determined. With SHA and declination, one can identify the body by reference to an almanac. Any method of solving a spherical triangle, with two sides and the included angle being given, is suitable for this purpose. "Short" methods such as H.O. Pubs. Nos. 208 and 211 include instructions for star identification by the tables provided. A large-scale, carefully drawn diagram on the plane of the celestial meridian, using the refinement shown in figure 22, should yield satisfactory results. Perhaps the simplest method of actual computation is by H.O. Pub. No. 214. Following the tables of computed altitude and azimuth for each latitude, a two-page star identification table is given.

IDENTIFICATION OF
NAVIGATIONAL STARS

Introduction—The following summary is not intended as a substitute for a star finder such as H.O. 2102-D, or of a knowledge of the heavens, but is given as a supplementary reference to assist in locating the 57 stars included in the main listing in the *Nautical Almanac*, plus Polaris. The observer is assumed to be at about the average latitude of the United States, unless another latitude is indicated. If a celestial body is said to be *east* of another, it is lower in the sky if both are rising and higher if both are setting. A body *north* of another is nearer the north celestial pole. Directions refer to great circles on the celestial sphere. Figures referred to are the star charts shown earlier, which should be of assistance in interpreting the descriptions given. It is assumed the reader is familiar with such well-known configurations as the big dipper and *Orion*. Constellation names are given in *italics*.

Acamar crosses the celestial meridian near the southern horizon during evening twilight in February, and during morning twilight in August. It is part of the constellation *Eridanus*, the river, which is not a striking configuration. It is the faintest star listed among the 57 in the almanac, but is the brightest in its immediate vicinity. The nearest bright star is Achernar, about 20° away in a generally southwesterly direction. Dec. 40°S, SHA 316°, mag. 3.1. Fig. 15.

Achernar, at the southern end of the inconspicuous

constellation *Eridanus,* the river, is one of the brightest stars of the southern hemisphere. It is not visible north of latitude 33°N. It crosses the celestial meridian during evening twilight in January, and during morning twilight in early August. Nearly a straight line is formed by Fomalhaut, about 40°WNW; Achernar; and Canopus, about the same distance in the opposite direction. However, since these stars are widely separated, the relationship is not striking. Achernar forms large triangles with Acamar and Ankaa, Ankaa and Al Na'ir, and with Al Na'ir and Peacock. Dec. 57°S, SHA 336°, mag. 0.6. Fig. 15.

Acrux is the brightest and most southerly star in the famed southern cross. It is not visible north of latitude 27°N. It crosses the celestial meridian during evening twilight in early June and during morning twilight in January. It is about 15°WSW of first magnitude Hadar and Rigil Kentaurus. Dec. 63°S, SHA 174°, mag. 1.1. Fig. 17.

Adhara. About 10°S and a little to the east of Sirius is a small, approximately equilateral triangle of three second magnitude stars. Adhara is the westernmost and brightest of the three. It crosses the celestial meridian to the south during evening twilight in March, and during morning twilight in October. Dec. 29°S, SHA 256°, mag. 1.6. Fig. 16.

Aldebaran. If the line formed by the belt of *Orion,* the hunter, is extended about 20° to the northwestward, and curved somewhat toward the north, it leads to first magnitude Aldebaran in *Taurus,* the bull. This is a group of stars forming a **V**. A long, curving line starting at Sirius extends through Procyon, Pollux, Capella, and Aldebaran. Dec. 16°N, SHA 292°, mag. 1.1. Fig. 16.

Alioth is the third star from the outer end of the handle of the big dipper, and the brightest star of the group. Dec. 56°N, SHA 167°, mag. 1.7. Fig. 17.

Alkaid is the star at the outer end of the handle of the big dipper, farthest from the bowl. It is the second brightest star of the group. Dec. 50°N, SHA 154°, mag. 1.9. Fig. 17.

Al Na'ir is the westernmost of two second magnitude stars of nearly equal brightness about midway between first magnitude Fomalhaut, approximately 20° to the northeast, and second magnitude Peacock, about the same distance in the opposite direction. A curved line extending eastward from the southern cross passes through Hadar and Rigil Kentaurus and, if extended with less curvature, leads first to Peacock and then to Al Na'ir. This star forms triangles with Fomalhaut and Ankaa, Ankaa and Achernar, and with Achernar and Peacock. It is not visible north of latitude 43°N. It crosses the celestial meridian during evening twilight early in December, and during morning twilight in June. Dec. 47°S, SHA 29°, mag. 2.2. Figs. 15, 18.

Alnilam is the middle star of the belt of *Orion,* the hunter. Dec. 1°S, SHA 277°, mag. 1.8. Fig. 16.

Alphard, a second magnitude star, is the brightest in the inconspicuous constellation *Hydra,* the water monster. The nearest bright star is first magnitude Regulus, about 20°NNE. It is about midway between the horizon and zenith when it crosses the celestial meridian to the southward during evening twilight in late April, and during morning twilight in November. Dec. 8°S, SHA 219°, mag. 2.2. Fig. 17.

Alphecca is the brightest star of *Corona Borealis,* the northern crown, about 20°ENE of first magnitude Arcturus. It forms a triangle with Arcturus and Alkaid. It crosses the celestial meridian near the zenith during evening twilight in July, and during morning twilight in February. Dec. 27°N, SHA 127°, mag. 2.3. Figs. 17, 18.

Alpheratz, a second magnitude star, is at the northeast corner of the great square of *Pegasus,* the winged horse, and is the brightest of the four stars forming the square. It crosses the celestial meridian near the zenith during evening twilight early in January, and during morning twilight in July. Dec. 29°N, SHA 359°, mag. 2.2. Fig. 15.

Altair is at the southern vertex of a large, nearly right triangle which is a conspicuous feature of the evening sky in late summer and in autumn. The right angle is at Vega and the northern vertex is at Deneb. All three are first magnitude stars. Two fainter stars close to Altair, one on each side in a line through Vega, form a characteristic pattern making Altair one of the easiest stars to identify. It crosses the celestial meridian during evening twilight in October, and during morning twilight in May. Dec. 9°N, SHA 63°, mag. 0.9. Fig. 18.

Ankaa, a second magnitude star, is the brightest star in inconspicuous *Phoenix.* It is surrounded by and forms a series of triangles with Diphda, Fomalhaut, Al Na'ir, Achernar, and Acamar. It crosses the celestial meridian low in the southern sky in January, and during morning twilight in July. Dec. 42°S, SHA 354°, mag. 2.4. Fig. 15.

Antares is the brightest star in the conspicuous constellation *Scorpio,* the scorpion, which is low in the southern sky during evening twilight in late July, and morning twilight in late February. No other first magnitude star is within 40° of Antares and none toward the north is within 60°. It has a noticeable reddish hue and in appearance somewhat resembles Mars, which is occasionally near it in the sky. Dec. 26°S, SHA 113°, mag. 1.2. Fig. 18.

Arcturus. The curved line along the stars forming the handle of the big dipper, if continued in a direction away from the bowl, passes through brilliant, first magnitude Arcturus. The distance from Alkaid, at the end of the big dipper, to Arcturus is a little more than the length of the dipper. Arcturus forms a large triangle with Alkaid and Alphecca. Dec. 19°N, SHA 147°, mag. 0.2. Figs. 17 and 18.

Atria is the brightest of three stars forming a small triangle called *Triangulum Australe,* the southern triangle, not far from the south celestial pole. It is not seen north of latitude 21°N. A line through the east-west arm of the southern cross, if continued toward the east and curved somewhat toward the south, leads first to Hadar, then to Rigil Kentaurus, then, by curving more sharply, to the northernmost star of the triangle, and finally to Atria, only about 21° from the south celestial pole. Dec. 69°S, SHA 109°, mag. 1.9. Fig. 17.

Avior is the westernmost star of *Vela,* the sails, or false southern cross, about 30°WNW of the true southern cross, about 15°ESE of the brilliant Canopus, and nearly enclosed within a large triangle formed by Canopus, Suhail, and Miaplacidus. It is not visible north of latitude 31°N. Below this, it crosses the celestial meridian low in the southern sky during evening twilight in April, and morning twilight in early November. Dec. 59°S, SHA 235°, mag. 1.7. Figs. 16 and 17.

Bellatrix is a second magnitude star north and a little west of the belt of *Orion,* the hunter. It is about equidistant from the belt and first magnitude, red Betelgeuse. Bellatrix is at the northwest corner of a box surrounding the belt of *Orion.* Dec. 6°N, SHA 279°, mag. 1.7. Fig. 16.

Betelgeuse is a conspicuous, reddish star of variable brightness about 10° north and a little east of the belt of *Orion,* the hunter. A line through the center of the belt and perpendicular to it passes close to red Betelgeuse to the north and blue Rigel about the same distance south of the belt. Betelgeuse and Rigel are at opposite corners of a box surrounding the belt of *Orion.* Dec. 7°N, SHA 272°, mag. 0.1–1.2 (variable). Fig. 16.

Canopus, second brightest star in the sky, is about 35° south of Sirius. A line extending eastward through the belt of *Orion* and curving toward the south passes first through Sirius, then through the small triangle of which Adhara is the brightest star, and finally to Canopus, which forms a large, almost equilateral triangle with Suhail and Miaplacidus. This triangle nearly encloses *Vela,* the sails or false southern cross, about 20°ESE of Canopus. Canopus is not visible north of latitude 37°N. It is on the edge of the Milky Way and while many relatively bright stars are nearby, none in the immediate vicinity of Canopus approaches it in brightness. Dec. 53°S, SHA 264°, mag. (−)0.9. Fig. 16.

79

Capella is a brilliant star about 45° north of the belt of *Orion,* the hunter. A curved line starting at Sirius and extending through Procyon, Pollux, Capella, Aldebaran, the belt of *Orion,* and back to Sirius forms an inverted tear-drop figure with Capella at the top and the various parts being about equally spaced along the curve. Capella crosses the celestial meridian near the zenith during evening twilight in early March, and during morning twilight in late September. Dec. 46°N, SHA 282°, mag. 0.2. Fig. 16.

Deneb is a bright star at the northeastern vertex of a large, nearly right triangle formed by Altair, Vega, and Deneb, the right angle being at Vega. These three stars are the brightest in the eastern sky during summer evenings. Deneb is not as bright as the other two, but is the brightest star in the constellation *Cygnus,* the swan. It crosses the celestial meridian near the zenith during evening twilight in November, and during morning twilight in late May. Dec. 45°N, SHA 50°, mag. 1.3. Fig. 18.

Denebola, in *Leo,* the lion, is a second magnitude star at the opposite end of the constellation from Regulus. A straight line from Regulus, on the west, to Arcturus, on the east, passes close to Denebola, which is somewhat nearer Regulus. Denebola crosses the celestial meridian to the south during evening twilight in May, and during morning twilight in December. Dec. 15°N, SHA 183°, mag. 2.2. Fig. 17.

Diphda. A line extending southward through the eastern side of the great square of *Pegasus,* the winged horse, and curving slightly toward the east, leads to second magnitude Diphda. The distance from the southern star of *Pegasus* to Diphda is about twice the length of one side of the square. Diphda is part of the inconspicuous constellation *Cetus,* the whale. The only nearby first magnitude star is Fomalhaut, about 25° in a generally southwest direction. Diphda, Fomalhaut, and Ankaa form a nearly equilateral triangle. Dec. 18°S, SHA 350°, mag. 2.2. Fig. 15.

Dubhe forms the outer rim of the bowl of the big dipper.

It and Merak (not one of the 57 navigational stars) are the two "pointers" used to locate Polaris, Dubhe being the one nearer the pole star. Dec. 62°N, SHA 195°, mag. 2.0. Fig. 17.

Elnath is a second magnitude star between Capella, about 15° to the north, and Betelgeuse, about 20° to the south. It is a little north of a line connecting Aldebaran and Pollux. It is at the end of the northern fork of V-shaped *Taurus,* the bull. Aldebaran is the principal star at the closed end of the V. This constellation is approximately 25°NNW of *Orion,* the hunter. Dec. 29°N, SHA 279°, mag. 1.8. Fig. 16.

Eltanin is the southernmost and brightest star in the inconspicuous constellation *Draco,* the dragon, south and somewhat east of the little dipper. A straight line extending northwestward through Altair and its two fainter companions passes first through brilliant Vega, and, about 15° beyond, to second magnitude Eltanin. Eltanin crosses the celestial meridian high in the sky toward the north during evening twilight in early September, and during morning twilight in late March. Dec. 51°N, SHA 91°, mag. 2.4. Fig. 18.

Enif is a third magnitude star approximately midway between Altair, about 25° west, and Markab, about 20°ENE. From Markab, at the southwestern corner of the great square of *Pegasus,* the winged horse, a line extending in a generally west-southwesterly direction passes through two almost equally spaced fourth magnitude stars. From the second of these, a line about 5° long extending in a northwesterly direction leads to Enif. Enif crosses the celestial meridian to the south during evening twilight in November, and during morning twilight in June. Dec. 10°N, SHA 35°, mag. 2.5. Figs. 15 and 18.

Fomalhaut is a first magnitude star well separated from stars of comparable brightness and from conspicuous configurations. A line through the western side of the great square of *Pegasus,* the winged horse, and extended about 45° toward the south passes close to Fomalhaut, which forms two large, nearly equilateral triangles with Diphda and Ankaa and with Ankaa and Al Na'ir. Dec. 30°S, SHA 16°, mag. 1.3. Fig. 15.

Gacrux is the northernmost star of the southern cross. It is bright for a second magnitude star, but its brilliance is overshadowed by the brighter *β Crucis* (not listed among the 57 navigational stars) and Acrux, the two brightest stars of the southern cross, and by Hadar and Rigil Kentaurus, about 15°ESE. Gacrux crosses the celestial meridian during evening twilight in early June, and during morning twilight in late December, but is not visible north of latitude 33°N. Dec. 57°S, SHA 173°, mag. 1.6. Fig. 17.

Gienah is a third magnitude star, the brightest in the constellation *Corvus,* the crow. A long, sweeping arc starting with the handle of the big dipper and extending successively through Arcturus and Spica leads to this relatively small, four-sided figure made up of third magnitude stars. Gienah is at the northwest corner. It crosses the celestial meridian during evening twilight in late May, and during morning twilight in December. Dec. 17°S, SHA 177°, mag. 2.8. Fig. 17.

Hadar is a first magnitude star about 10° east of the southern cross, and about 5° west of Rigil Kentaurus, the brightest of several bright stars in this part of the sky. Dec. 60°S, SHA 150°, mag. 0.9. Fig. 17.

Hamal is the brightest star of the inconspicuous constellation *Aries,* the ram. A line through the center of the great square of *Pegasus,* the winged horse, extended about 25° east, and curved slightly toward the north, leads to Hamal. It is over the meridian to the south during evening twilight in January, and during morning twilight in August. Dec. 23°N, SHA 329°, mag. 2.2. Fig. 15.

Kaus Australis is near the southern end of a group of second and third magnitude stars forming the constellation *Sagittarius,* the archer, about 25°ESE of Antares, in *Scorpio,* the scorpion. It is about 10°SW of Nunki, also in *Sagittarius,* and about the same distance ENE of Shaula, in *Scorpio.* With Antares, Sabik, and Nunki, it forms a large, poorly defined box. It is over the meridian to the south during evening twilight in September and during morning twilight in April. Dec. 34°S, SHA 85°, mag. 2.0. Fig. 18.

Kochab forms the outer rim of the bowl of the little dipper, at the opposite end from Polaris, about 15° north. It is directly above the pole during evening twilight in early July and during morning twilight in January; and directly below the pole, low in the northern sky, during evening twilight of early February and morning twilight of late August. Dec. 74°N, SHA 137°, mag. 2.2. Fig. 18.

Markab is the star at the southwest corner of the great square of *Pegasus,* the winged horse, at the opposite corner from Alpheratz. It is over the celestial meridian to the south during evening twilight in December, and during morning twilight late in June. Dec. 15°N, SHA 14°, mag. 2.6. Fig. 15.

Menkar is a third magnitude star at the eastern end of the inconspicuous constellation *Cetus,* the whale. No bright stars are nearby. A straight line from Aldebaran extending about 25° in the direction indicated by the point of the V of *Taurus,* the bull, leads to Menkar. A long, straight line from Fomalhaut east-northeastward through Diphda, and extended about 40°, leads to Menkar. It crosses the celestial meridian during evening twilight in February, and during morning twilight in August. Dec. 4°N, SHA 315°, mag. 2.8. Figs. 15 and 16.

Menkent is a second magnitude star about 25° north of Hadar and about 30° northeast of the southern cross. A line from Gienah across the opposite corner of the small, four-sided *Corvus,* the crow, and then curving a little toward the east, leads to Menkent. A number of third magnitude stars are nearby, but they do not form a conspicuous configuration. With Antares and Rigil Kentaurus, Menkent forms a large triangle. It crosses the celestial meridian low in the southern sky during evening twilight in late June and during morning twilight in early January. Dec. 36°S, SHA 149°, mag. 2.3. Figs. 17 and 18.

Miaplacidus is a second magnitude star about 10° south of the false southern cross. It is the nearest of the 57 navigational stars to the south celestial pole, about 20° away, and is not visible north of latitude 20° N. With Suhail and brilliant Canopus it forms a large, nearly equilateral triangle almost enclosing the false southern cross. South of latitude 20°S, it does not set, but circles the south celestial pole in a clockwise direction, reaching its maximum altitude above the pole during evening twilight in early May and during morning twilight in November. Dec. 70°S, SHA 222°, mag. 1.8. Figs. 16 and 17.

Mirfak is a second magnitude star at the northeastern end of a gently curving line extending in a northeasterly direction from Alpheratz at the northeastern corner of the great square of *Pegasus,* the winged horse, through two other second magnitude stars, Mirach and Almach, not included among the 57 navigational stars. Mirfak is about 25° east and a little south of *Cassiopeia,* and about 20°WNW of Capella. A line from Kochab through Polaris, and curved slightly toward the east, leads to Mirfak. Dec. 50°N, SHA 310°, mag. 1.9. Figs. 15 and 16.

Nunki is the more northerly of the two brightest stars of a group of second and third magnitude stars forming the constellation *Sagittarius,* the archer, about 30°E of Antares. It is about 10° NE of Kaus Australis, also in *Sagittarius.* With Sabik, Antares, and Kaus Australis, it forms a large, poorly defined box. It is over the meridian to the south during evening twilight in early October and during morning twilight in April. Dec. 26°S, SHA 77°, mag. 2.1. Fig. 18.

Peacock, the brightest star in the southern constellation of the same name, is not a part of a conspicuous configuration of stars. A curved line extending eastward from the southern cross passes through Hadar and Rigil Kentaurus and, if extended with less curvature, leads to Peacock, about 30° southeast of *Scorpio,* the scorpion, and about 20° southwest of Al Na'ir. With Al Na'ir and Achernar it forms a large, poorly defined triangle. It crosses the celestial meridian during evening twilight in early November, and during morning twilight in late May, but is not visible north of latitude 33°N. Dec. 57°S, SHA

54°, mag. 2.1. Figs. 15 and 18.

Polaris is not listed among the 57 navigational stars, but is treated separately because it is less than 1° from the north celestial pole. It is about midway between the big dipper and *Cassiopeia*. A line through Dubhe and Merak (not one of the 57 navigational stars), the pointers forming the outer side of the bowl of the big dipper, if extended northward for about 30°, leads almost directly to Polaris. A line extending north from Alpheratz at the northwest corner of the great square of *Pegasus,* the winged horse, passes through Caph (not one of the 57 navigational stars) in *Cassiopeia* and then Polaris at about equal intervals. Dec. 89°N, SHA 332°, mag. 2.1. Figs. 15-18.

Pollux is the brighter of the "twins of *Gemini,*" two relatively bright stars about 45°NE of *Orion,* the hunter, and about 45°ENE of Aldebaran. A curved line starting at Sirius extends through Procyon, Pollux, and Capella, all first magnitude stars. Dec. 28°N, SHA 244°, mag. 1.2. Fig. 16.

Procyon is a bright star about 30° east of *Orion,* the hunter. A curved line starting at Sirius extends through Procyon, Pollux, and Capella, all first magnitude stars. Dec. 5°N, SHA 246°, mag. 0.5. Fig. 16.

Rasalhague forms a large, nearly equilateral triangle with Altair and Vega, Rasalhague being at the western vertex. Both of the other stars are considerably brighter than Rasalhague. It crosses the celestial meridian to the south during evening twilight in early September, and during morning twilight in late March. Dec. 13°N, SHA 97°, mag. 2.1. Fig. 18.

Regulus is at the opposite end of *Leo,* the lion, from Denebola, and is the brightest star of the constellation. A line through Dubhe and Merak (not one of the 57 navigational stars), the pointers by which Polaris is usually identified, extended about 45° *southward,* and curved slightly toward the west, leads to Regulus, which forms the southern end of the handle of the sickle, part of *Leo.* Dec. 12°N, SHA 209°, mag. 1.3. Fig. 17.

Rigel is a brilliant bluish star about 10°S and a little to the west of the belt of *Orion,* the hunter. A line through the center of the belt and perpendicular to it passes close to blue Rigel to the south and red Betelgeuse about the same distance north of the belt. Rigel and Betelgeuse are at opposite corners of a box surrounding the belt of *Orion.* Dec. 8°S, SHA 282°, mag. 0.3. Fig. 16.

Rigil Kentaurus is the brighter and more easterly of two first magnitude stars about 15° east of the southern cross. It is over the meridian during evening twilight in early July, and during morning twilight in late January, but is not visible north of latitude 29°N. Dec. 61°S, SHA 141°, mag. 0.1. Figs. 17 and 18.

Sabik is part of the inconspicuous constellation *Ophiuchus,* the serpent holder, about 20° north of *Scorpio,* the scorpion. With Antares, Kaus Australis, and Nunki, it forms a large, poorly defined box in the southern sky on summer evenings. Sabik crosses the celestial meridian during evening twilight in August, and during morning twilight in March. Dec. 16°S, SHA 103°, mag. 2.6. Fig. 18.

Schedar is the southernmost star of the W (or M) of *Cassiopeia,* on the opposite side of Polaris from the big dipper. It is the second star from the leading edge of this configuration as it circles the north celestial pole. Dec. 56°N, SHA 351°, mag. 2.5. Figs. 15, 16, and 18.

Shaula is a second magnitude star marking the end of the tail of *Scorpio,* the scorpion, at the opposite end from Antares. This constellation is low in the southern sky on summer evenings. Shaula is about 15° southeast of Antares and about 10°WSW of Kaus Australis. It crosses the celestial meridian during evening twilight in early September, and during morning twilight in March. Dec. 37°S, SHA 97°, mag. 1.7. Fig. 18.

Sirius, the brightest star in the heavens, is in the constellation *Canis Major,* the "large dog" of *Orion,* the hunter. The line formed by the belt of *Orion,* if extended about 20° to the eastward and curved toward the south, leads to Sirius. Dec. 17°S, SHA 259°, mag. (−)1.6. Fig. 16.

Spica is the brightest star of *Virgo,* the virgin, an inconspicuous constellation on the celestial equator to the south during evening twilight in early summer. The curved line along the stars forming the handle of the big dipper, if continued in a direction away from the pointers, passes through Arcturus and then Spica. The distance between Alkaid, at the end of the big dipper, and Arcturus is about the same as that between Arcturus and Spica, and is a little more than the length of the big dipper. Spica crosses the celestial meridian during evening twilight in June, and during morning twilight late in December. Dec. 11°S, SHA 159°, mag. 1.2. Fig. 17.

Suhail is one of a number of second magnitude stars extending along the Milky Way between Sirius and the southern cross. It is about 10° north of the false southern cross, which is nearly enclosed by a large, almost equilateral triangle formed by Suhail, Canopus, and Miaplacidus. Canopus and Suhail are on opposite edges of the Milky Way, with a number of second magnitude stars between them. A straight line extending eastward through the east-west arm of the southern cross leads to Suhail, about 35° away. In the southern United States, Suhail crosses the celestial meridian near the southern horizon during evening twilight in April, and during morning twilight in November. Dec. 43°S, SHA 223°, mag. 2.2. Figs. 16 and 17.

Vega is the brightest star north ot tne celestial equator, and the third brightest in the entire sky. It is at the western vertex and the nearly right angle of a large triangle which is a conspicuous feature of the evening sky in late summer and in autumn. The other two stars of the triangle are Altair and Deneb, both of the first magnitude. Vega passes through the zenith approximately at latitude 38°45′N during evening twilight in September and during morning twilight in April. Dec. 39°N, SHA 81°, mag. 0.1. Fig. 18.

Zubenelgenubi, a third magnitude star, is the southern (or western) basket of *Libra,* the balance. The boxlike *Libra* is about 25°WNW of Antares, in *Scorpio,* the scorpion. A long line extending eastward from Alphard, between Gienah and Spica, leads to Zubenelgenubi. Dec. 16°S, SHA 138°, mag. 2.9. Figs. 17 and 18.

APPENDIX A

NAVIGATIONAL STARS AND THE PLANETS

Name	Pronunciation	Bayer name	Origin of name	Meaning of name	Distance*
Acamar	ā′kȧ·mär	θ Eridani	Arabic	another form of Achernar	120
Achernar	ā′kẽr·när	α Eridani	Arabic	end of the river (Eridanus)	72
Acrux	ā′krŭks	α Crucis	Modern	coined from Bayer name	220
Adhara	ȧ·dä′rȧ	ε Canis Majoris	Arabic	the virgin(s)	350
Aldebaran	ăl děb′ȧ·rȧn	α Tauri	Arabic	follower (of the Pleiades)	64
Alioth	ăl′ĭ·ŏth	ε Ursa Majoris	Arabic	another form of Capella	49
Alkaid	ăl·kād′	η Ursa Majoris	Arabic	leader of the daughters of the bier	190
Al Na'ir	ȧl·när′	α Gruis	Arabic	bright one (of the fish's tail)	90
Alnilam	ăl′nĭ·lăm	ε Orionis	Arabic	string of pearls	410
Alphard	ăl′färd	α Hydrae	Arabic	solitary star of the serpent	200
Alphecca	ăl·fĕk′ȧ	α Corona Borealis	Arabic	feeble one (in the crown)	76
Alpheratz	ăl·fē′răts	α Andromedae	Arabic	the horse's navel	120
Altair	ăl·târ′	α Aquilae	Arabic	flying eagle or vulture	16
Ankaa	ăn′kä	α Phoenicis	Arabic	coined name	93
Antares	ăn·tä′rēz	α Scorpii	Greek	rival of Mars (in color)	250
Arcturus	ärk·tū′rŭs	α Boötis	Greek	the bear's guard	37
Atria	ät′rĭ·ȧ	α Trianguli Australis	Modern	coined from Bayer name	130
Avior	ā′vĭ·ôr	ε Carinae	Modern	coined name	350
Bellatrix	bĕ·lā′trĭks .	γ Orionis	Latin	female warrior	250
Betelgeuse	bĕt′ĕl·jū z	α Orionis	Arabic	the arm pit (of Orion)	300
Canopus	kȧ·nō′pŭs	α Carinae	Greek	city of ancient Egypt	230
Capella	kȧ·pĕl′ȧ	α Aurigae	Latin	little she-goat	46
Deneb	dĕn′ĕb	α Cygni	Arabic	tail of the hen	600
Denebola	dē·nĕb′ō·lȧ	β Leonis	Arabic	tail of the lion	42
Diphda	dĭf′dȧ	β Ceti	Arabic	the second frog (Fomalhaut was once the first)	57
Dubhe	dŭb′ē	α Ursa Majoris	Arabic	the bear's back	100
Elnath	ĕl′năth	β Tauri	Arabic	one butting with horns	130
Eltanin	ĕl·tä′nĭn	γ Draconis	Arabic	head of the dragon	150
Enif	ĕn′ĭf	ε Pegasi	Arabic	nose of the horse	250
Fomalhaut	fō′mȧl·ŏt	α Piscis Austrini	Arabic	mouth of the southern fish	23
Gacrux	gā′krŭks	γ Crucis	Modern	coined from Bayer name	72
Gienah	jē′nȧ	γ Corvi	Arabic	right wing of the raven	136
Hadar	hä′där	β Centauri	Modern	leg of the centaur	200
Hamal	hăm′ăl	α Arietis	Arabic	full-grown lamb	76
Kaus Australis	kôs ôs·trā′lĭs	ε Sagittarii	Ar., L.	southern part of the bow	163
Kochab	kō′kăb	β Ursa Minoris	Arabic	shortened form of "north star" (named when it was that, c. 1500 BC–AD 300)	100
Markab	mär′kăb	α Pegasi	Arabic	saddle (of Pegasus)	100
Menkar	mĕn′kär	α Ceti	Arabic	nose (of the whale)	1, 100
Menkent	mĕn′kĕnt	θ Centauri	Modern	shoulder of the centaur	55
Miaplacidus	mī′ȧ·plăs′ĭ·dŭs	β Carinae	Ar. L.	quiet or still waters	86
Mirfak	mĭr′făk	α Persei	Arabic	elbow of the Pleiades	130
Nunki	nŭn′kē	σ Sagittarii	Bab.	constellation of the holy city (Eridu)	150
Peacock	pē′kŏk	α Pavonis	Modern	coined from English name of constellation	250
Polaris	pô·lā′rĭs	α Ursa Minoris	Latin	the pole (star)	450
Pollux	pŏl′ŭks	β Geminorum	Latin	Zeus' other twin son (Castor, α Geminorum, is first twin)	33
Procyon	prō′sĭ·ŏn	α Canis Minoris	Greek	before the dog (rising before the dog star, Sirius)	11
Rasalhague	räs′ȧl·hā′gwē	α Ophiuchi	Arabic	head of the serpent charmer	67
Regulus	rĕg′ū·lŭs	α Leonis	Latin	the prince	67
Rigel	rī′jĕl	β Orionis	Arabic	foot (left foot of Orion)	500
Rigil Kentaurus	rī′jĭl kĕn·tô′rŭs	α Centauri	Arabic	foot of the centaur	4. 3
Sabik	sä′bĭk	η Ophiuchi	Arabic	second winner or conqueror	69
Schedar	shĕd′är	α Cassiopeiae	Arabic	the breast (of Cassiopeia)	360
Shaula	shō′lä	λ Scorpii	Arabic	cocked-up part of the scorpion's tail	200
Sirius	sĭr′ĭ·ŭs	α Canis Majoris	Greek	the scorching one (popularly, the dog star)	8. 6
Spica	spī′kȧ	α Virginis	Latin	the ear of corn	155
Suhail	sōō·hāl′	λ Velorum	Arabic	shortened form of Al Suhail, one Arabic name for Canopus	200
Vega	vē′gȧ	α Lyrae	Arabic	the falling eagle or vulture	27
Zubenelgenubi	zōō·bĕn′ĕl·jē·nū′bē	α Librae	Arabic	southern claw (of the scorpion)	66

PLANETS

Name	Pronunciation	Origin of name	Meaning of name
Mercury	mûr′kū·rĭ	Latin	god of commerce and gain
Venus	vē′nŭs	Latin	goddess of love
Earth	ûrth	Mid. Eng.	—
Mars	märz	Latin	god of war
Jupiter	jōō′pĭ·tẽr	Latin	god of the heavens, identified with the Greek Zeus, chief of the Olympian gods
Saturn	săt′ẽrn	Latin	god of seed-sowing
Uranus	ū′rȧ·nŭs	Greek	the personification of heaven
Neptune	nĕp′tūn	Latin.	god of the sea
Pluto	plōō′tō	Greek	god of the lower world (Hades)

Guide to pronunciations:
fāte, ădd, fīnȧl, lȧst, ȧbound, ärm; bē, ĕnd, camĕl, readẽr; īce, bĭt, anĭmal; ōver, pȯetic, hŏt, lôrd, mōōn; tūbe, ûnite, tŭb, circ⊍s, ûrn

*Distances in light-years. One light-year equals approximately 63,300 A U, or 5,880,000,000,000 miles. Authorities differ on distances of the stars; the values given are representative.

APPENDIX B

A Catalog of Marine Sextants

Marine sextants range in price and precision from modest through moderate to expensive; and from good to better to best. As is true with most precision instruments, you pay your money and you take your choice—and you usually get what you pay for. Also, as with most precision instruments, used sextants *are* available, but they are usually hard to find and you may wait years before you find one that suits both your nautical needs and your pocketbook.

Yachtsmen have circumnavigated the globe with plastic sextants and ordinary watches or clocks; Joshua Slocum managed the same feat with a used sextant and an old alarm clock. But most of us would prefer something better—if not *the best,* then the best that we can afford.

The following section is not a total catalog of all the sextants that are available; nor is it intended to prove that one sextant is better than another, or that *this* instrument is a better buy than *that* instrument. Instead, it is meant to serve as a starting point—a new buyer's guide to a number of excellent marine sextants. Most are available from the major marine stores or distributors, often through catalogs. Excellent catalogs, listing a variety of sextants, are distributed by firms and navigation schools such as Kleid Navigation, Inc.; Coast Navigation School; Nautech Maritime; and Weems & Plath, Inc. Usually, however, it is preferable to see before you buy, for the "feel" of a good sextant is a key point in its use. It is also a satisfying experience!

In presenting the following selection of sextants, the Publisher has drawn from recent catalog copy used by the various manufacturers and distributors.

DAVIS MARINE SEXTANT

Made of heavy gauge plastic with a clearly marked vernier, this practice sextant is sufficiently accurate for taking sights at sea. It is also good for horizontal angles. The Davis is an excellent instrument in the economy price range, and is useful as a second sextant.

EBBCO MICROMETER SEXTANT

This fine instrument—the EBBCO—is made in England of precision-molded makrolon, a very stable plastic. It differs from most inexpensive sextants in that it has all the features of the modern quality micrometer sextant that contribute to the ease of sight taking. It has a 2-power telescope, a micrometer drum with an endless tangent screw, a 0.2 minute vernier for drum, highly visible markings on arc, drum and vernier, and multiple index and horizon shades which have been tested to assure protection from ultraviolet rays. It is also a good instrument for practice and for use as a second sextant for offshore cruising or racing boats.

DAVIS MARK 12 SEXTANT

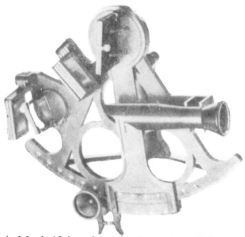

The Davis Mark 12 is a fine plastic sextant. It has a new 3-power metal telescope with ground-glass lenses. The Mark 12's distinct white markings on a black background make the 2′ vernier quick and easy to read. A smooth-acting, fine-adjusting knob on the index arm works in conjunction with the rapid motion clamp. Within the limitations of plastic, the Mark 12 is very well constructed. It has standard mirror adjustments, with thumb screws, and large mirrors that are fully protected. Its horizon mirror has both silvered and clear halves, and it has a full complement of seven sun shades.

SIMEX MARK I SEXTANT/SIMEX JUNIOR SEXTANT

A classic sextant and one of the finest, the Simex Mark I, above left, is a traditional instrument with optical shade glasses plus all the modern refinements. The MARK I has a 4 × 40mm coated optics telescope, astigmatizer lens, 33mm × 49mm aluminized index and 28mm × 35mm aluminized horizon mirrors.

The compact Simex Junior, above right, is a fully professional instrument for the small-boat navigator. This three-quarter-size sextant features a full-size micrometer drum and handle, and is built to rigid specifications that give it precision capabilities. It has a 3 × 26mm coated optics telescope, 22mm × 36mm aluminized index mirror, 22mm × 29mm aluminized horizon mirror, and optical dyed shade glasses. Its brass arc reads from –5° to 106°.

This precision, two-scope sextant is the result of extensive testing of various combinations of optics, filters, and mirrors under actual sea conditions. It offers the navigator great flexibility and accuracy in dealing with the demanding conditions encountered at sea. This precision instrument—with all its advanced features—is medium-priced.

The Nautech Master's 7-power scope, with its excellent field of vision, provides accuracy and ease of taking sights from a rolling deck. In addition, there's the Master's 12-power collimation scope and a 4 × 40 star scope, providing a combination of optics for every purpose; 7 large optical quality shade filters for a wide range of filter capability for use under cloudy or hazy conditions; black enamel arcs and white numerals for easy reading under limited light conditions; non-tarnishing and lighted micrometer drums and arcs for improving the speed and accuracy of taking twilight sights. Reads to 10 seconds of arc. The Master Sextant is available with 7 × 35 scope, 4 × 40 and 12 × 20 scopes, 7 × 35 and 12 × 20 scopes, or 4 × 40, 7 × 35 and 12 × 20 scopes.

NAUTECH GEMINI

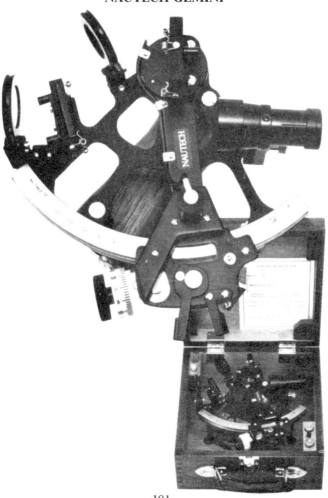

For the occasional or apprentice navigator, the Gemini ¾ scale sextant is a fine, practical sextant to acquire the navigation skills needed at sea. The Gemini is both rugged and has a full-size illuminated micrometer drum and angled teak handle. It has excellent precision capabilities when used from a stable platform. It is an effective training or "backup" instrument, but because of its smaller-sized mirrors and optics, it is not suitable for serious, offshore work. The Gemini has a 3 × 26mm (Field of View - 8°) telescope and all optics are fully coated. It has a 23 × 37mm index mirror, 23 × 37mm horizon mirror, and optical dyed glass for the seven shade filters.

INTERNATIONAL NAUTICAL SEXTANTS

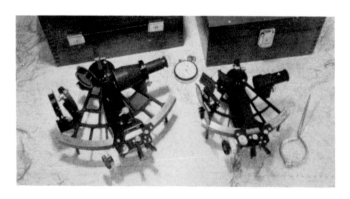

The International Nautical Sextants are manufactured in Japan to Weems & Plath specifications and are available in the full size "Professional" and the smaller "Mini-sextant" designs. Produced in large quantities to sell in the medium price range, these sextants have crossed many oceans in the hands of competent navigators and have been widely used by students of celestial navigation.

The full size professional model, which is available with

either a 4 × 40 telescope or a 7 × 35 monocular, is equipped with optically ground individual shades of neutral tint or variable density polarized shades, and has a light on both the arc and the micrometer drum. Two penlight batteries are contained in the handle. The mini-sextant, which is unlighted and weighs only 2 lbs. 1 oz., is designed for the yachtsman who wants a small but rugged sextant that does not take up much storage space and is more easily handled aboard a yacht at sea. Given reasonable care, either model should last a lifetime.

THE CAPELLA SERIES SEXTANTS

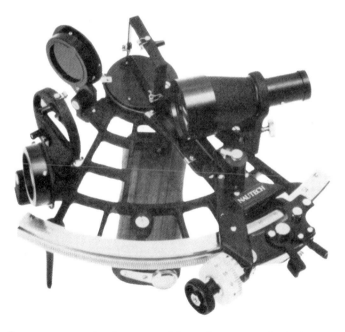

Improvements in these new Nautech sextants include an inclined teakwood handle that affords greater comfort as well as enhancing the sextant's appearance. The handle houses two penlight batteries that provide illumination to both the arc and drum. The Capella I series comes with a 4 × 40 scope with a field of view of 6°.

The Capella II provides the same features as the Capella I but comes with a 7 × 35mm prismatic monocular as standard with a field of view of 6.5° and a better relative light efficiency factor. Both instruments have a milk-white plastic micrometer drum with one-minute scale and vernier that reads to 0.2 of minute.

HEATH "HEZZANITH" SEXTANT - MODEL 130M

Heath has been making fine instruments for 120 years. The company's Model 130M Sextant represents a good dollar value. Its design incorporates the basic features of more expensive instruments, and it has proven itself over many years of successful use by professionals and yachtsmen. A unique feature of Heath sextants is that the rack teeth are cut on the back side of the limb, thereby reducing radial pressure (and wear) on the pivot arm, and increasing chances for long-term accuracy. Its specifications include a divided arc from –5° to 125°, 44.5mm × 33.3mm index and 33.3mm × 31.8mm horizon mirrors, and a full set of seven special sextant-neutral shade glasses. The telescope is 3 × 28.5.

KELVIN HUGHES MATE SEXTANT

This precision micrometer sextant is made by the well-known British firm of Kelvin Hughes, which has for years manufactured navigation instruments for the British merchant marine and navy. It is a full-sized quality sextant that is used by yachtsmen and others who want a professional sextant from a reliable manufacturer. It has the distinctive 3-circle design with pebble grey finish, large micrometer drum, and a mahogany handle. Its specifications include an arc from –5° to +125° on a satin chrome-plated limb, vernier divided in 0.2 minutes, rectangular, protected, 1.3 inches × 1.9 inches mirrors, and 4 index and 3 horizon shades in neutral tint and 1 dark telescope shade, and a telescope with achromatic erect star, magnification 2.5, field 7°, and aperture 1.22 inches.

THE "FREIBERGER" DRUM SEXTANT

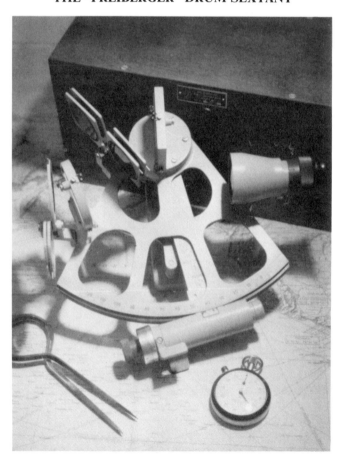

Well known by yachtsmen in the United Kingdom and Europe, the "Freiberger" Drum Sextant has been built to exacting German specifications. It is now available for the first time in the United States. Weems & Plath, the U.S. distributor, added this sextant to fill a price gap. The "Freiberger" is a full-sized modern sextant that weighs only 3 lbs. It has a fully enclosed drum clamping device and utilizes large mirrors (which are interchangeable with Plath) to retain excellent light intensity and field of vision. It is equipped with a 3½ × 40 telescope. The "Freiberger" contains no error greater than 40″ for any position on the arc.

SIMEX MARINER SEXTANT

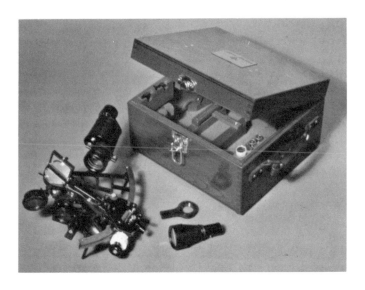

A precision instrument, engineered for accuracy, the Simex Mariner is made to the specifications of Captain Svend T. Simonsen, founder of Coast Navigation School.

The Mariner sextant, which is equipped with two scopes and will give a vernier reading to within a tenth of a minute, has the following specifications: prism monocular 7 × 35mm and 4 × 40mm coated optics telescopes, 33mm × 49mm aluminized index and 50mm circular aluminized horizon mirrors, and an arc that reads from –5° to +125°.

THE NAUTECH MODEL 733

The new Nautech professional Model 733 Sextant incorporates all the design features requested at the 4th Annual Navigation Symposium at the United States Naval Academy in 1975.

Its large mirrors provide maximum light-gathering power for twilight sights. The index mirror is 57 × 42mm and the horizon mirror is a 57mm. The sextant is available with a choice of either a 4 × 40 telescope having relative light efficiency value of 75% for star sights or a 6 × 30 prismatic monocular which gives a wider field of view and higher resolving power for sun sights and horizontal and vertical bearing angles. The 6 × 30 monocular has a relative light efficiency value of 38%. Seven optical shades are provided, and an extra polaroid glass and shade glass is provided for the eyepiece of the telescope. It is accurate to 10 seconds of arc and the sextant's vernier reads to .2 minutes of arc. Full illumination of both the arc and drum are provided.

HEATH "HEZZANITH" SEXTANT - MODEL MC1

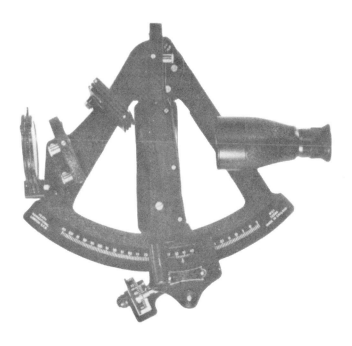

Heath's finest professional sextant, the Model MC1, provides a screw-focus 4-power telescope, and the necessary large and precise optics to go with it. As with the 130M, its rack is cut on the backside of the limb for long-term accuracy. The MC1 with its special aluminum alloy frame, is the lightest of professional sextants. Its specifications include: an arc divided from –5° to 125°, a vernier that reads to 0.2′, 54mm × 40mm index and 57mm diameter horizon mirrors, a full set of seven large shade glasses, and a 4 × 40 screw focusing telescope.

CASSENS & PLATH SEXTANT

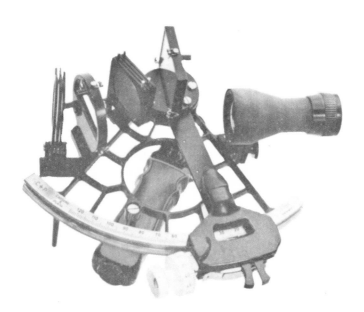

The Cassens & Plath micrometer drum sextant has world-wide recognition as a fine precision instrument. The large high-quality optical system facilitates good twilight sights.

Supplied in special brass alloy or lightweight aluminum alloy, every Cassens & Plath sextant has a guaranteed accuracy of the arc of better than 10 seconds. Its specifications include an arc that is divided from –5° to 125°, a vernier that reads to 0.2′, 56mm × 42mm index and 57mm diameter horizon mirrors, a full set of neutral shade glasses, and 4 × 40 or 6 × 30 telescopes.

U.S. NAVY MARK III MOD 1 SEXTANT

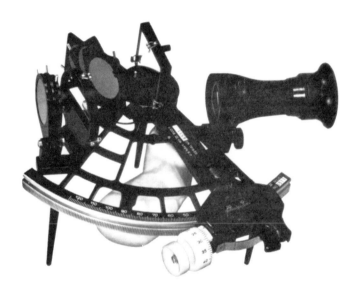

This instrument is the successor to the famous World War II U.S. Navy Mark II Sextant. It is of improved design with a guaranteed accuracy of 8 seconds of arc. Its newly designed handle houses two penlight batteries which provide illumination to the arc and drum.

The mirror sizes are similar to the Nautech Model 733 Sextant. It has neutral-density (not polarized) filters and shades. The vernier reads to 0.2 minutes of arc. This sextant is built in accordance with Navy DWG specifications. It comes with spare mirrors, batteries, and a 4 × 40 scope with a rubber eyepiece.

THE WEEMS & PLATH SEXTANT

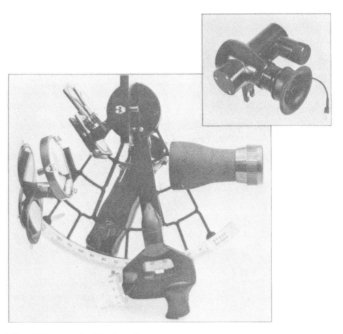

The Weems & Plath Micrometer Drum Sextant is considered by many navigators to be the finest instrument available throughout the maritime world. Manufactured in Germany by Weems & Plath, the sextant has a guaranteed accuracy of arc of better than 10 seconds of arc. The frame and the arc are constructed of a special brass alloy and the handle is slightly inclined to provide a natural grip. Total weight with 4 × 40 telescope is 4 lbs. The mirrors are extra large and are of the finest optical quality, providing a high light intensity. The horizon mirror is 50 millimeters in diameter. For daytime observations sunshades are provided in varying densities of neutral shades. An astigmatizer lens can be added if desired. A vernier on the micrometer drum provides a readout of two tenths minutes of arc. Lighting of the arc and micrometer drum is provided by two standard U.S. penlight batteries in the handle.

The sextants are available with a choice of 4 × 40 or 6 × 30 telescopes, each with a silicone rubber eyeshield. A lighted bubble attachment (inset) for making night observations is also available. For those who desire less weight, this sextant is also available with a frame and arc made of lightweight aluminum alloy. Total weight with a 4 × 40 telescope is only 3¼ lbs. The alloy Weems & Plath sextant has been treated and enameled to prevent corrosion. The accuracy and other specifications of the lightweight model are the same as the standard brass sextant.